SO YOU CREATED A WORMHOLE

SO YOU CREATED A WORMHOLE

THE TIME TRAVELER'S GUIDE TO TIME TRAVEL

PHIL HORNSHAW & NICK HURWITCH

B

BERKLEY BOOKS, NEW YORK

BERKLEY BOOKS
Published by the Penguin Group
Penguin Group (USA) Inc.
375 Hudson Street, New York, New York 10014, USA
Penguin Group (Canada), 90 Eglinton Avenue East, Suite 700, Toronto, Ontario M4P 2Y3,
Canada (a division of Pearson Penguin Canada Inc.) • Penguin Books Ltd., 80 Strand,
London WC2R 0RL, England • Penguin Group Ireland, 25 St. Stephen's Green, Dublin 2,
Ireland (a division of Penguin Books Ltd.) • Penguin Group (Australia), 250 Camberwell
Road, Camberwell, Victoria 3124, Australia (a division of Pearson Australia Group Pty.
Ltd.) • Penguin Books India Pvt. Ltd., 11 Community Centre, Panchsheel Park, New
Delhi—110 017, India • Penguin Group (NZ), 67 Apollo Drive, Rosedale, Auckland 0632,
New Zealand (a division of Pearson New Zealand Ltd.) • Penguin Books (South Africa)
(Pty.) Ltd., 24 Sturdee Avenue, Rosebank, Johannesburg 2196, South Africa

Penguin Books Ltd., Registered Offices: 80 Strand, London WC2R 0RL, England

The publisher does not have any control over and does not assume any responsibility for
author or third-party websites or their content.

PUBLISHING HISTORY
Berkley trade paperback edition / April 2012

Library of Congress Cataloging-in-Pubication Data

Hornshaw, Phil.
So you created a wormhole : the time traveler's guide to time travel /
Phil Hornshaw and Nick Hurwitch.
p. cm.
ISBN 978-0-425-24558-3
1. Science—Humor. 2. Time travel in literature. 3. Space and time.
I. Hurwitch, Nick. II. Title.
PN6231.S4H67 2012
818'.602—dc23
2011038455

PRINTED IN THE UNITED STATES OF AMERICA

10 9 8 7 6 5 4 3 2 1

ALWAYS LEARNING   PEARSON

ACKNOWLEDGMENTS

Brandi Bowles, a kicker of many asses.

Dr. Emmett Brown, whose brain damage made this all possible.

James Cameron, a visionary of horrific robot futures.

Albert Einstein, an incomparable genius.

iam8bit (Nick Ahrens, Jon M. Gibson, Taylor Harrington, Amanda White), for their fearless support of all things nerdy.

Jenni Frisbie, for believing in our book.

Bob Gale, for his contributions to our childhoods and for keeping Marty from fading away.

Brian Greene, whose writing and Radiolab appearances made us smarter (we think).

Alex Griendling, for his killer 2011 Time Travel Calendar.

Stephen Hawking, for spaghettification and using his brain powers for good.

Aled Lewis, for converting our vague instructions into incredible illustrations.

Brandy Rivers, for her infinite enthusiasm.

Steven Spielberg, for all those gifts of imagination.

H.G. Wells, for inventing the luxury time machine.

Robert Zemeckis, whose brilliant movies helped shape our futures.

Caitlin M. Foyt, whose laughter and encouragement sustained us.

Amanda White, whose support and reasoned advice calmed many would-be freakouts.

Our legion of friends-turned-editors, who helped when self-assurance faltered and typos threatened to overthrow us.

Berkley/Penguin, and its team of do-gooders who are helping to save time traveler lives in the name of science:

Andie Avila, a phenomenal editor who managed to make sense of all (or most) of what we wrote.

Pauline Neuwirth, the book designer who fit together scraps of burned paper from us to make the thing you're holding in your hands.

Tiffany Estreicher, for helping convey whatever we were gesturing wildly about into reality.

Pam Barricklow, who made sure it all went smoothly.

Erica Martirano, for the management of marketing, and therefore, the saving of less well-connected time travelers.

Rosanne Romanello and Jodi Rusoff, for their excitement and dedication as publicists.

Diana Kolsky, for creating a spectacular cover despite our constant meddling.

And finally,

Ricky the Intern, for his brave sacrifices in testing lots of this stuff to see if it would work. You'll always be remembered, Randy.

SO YOU CREATED A WORMHOLE

DO NOT READ THIS BOOK

WHAT YOU HAVE here is *The Time Traveler's Guide to Time Travel*. Nine times out of ten, that makes you a time traveler.

We know what you're thinking: "Um, I'm not a time traveler, I'm just interested in this robust and well-written piece of literature." WRONG. This book is the world's first and only field manual to time travel; ergo, it spends most of its existence traveling through time.

And because a guide to time travel is so very likely to have traveled through time before, the authors have deduced that nine times out of ten, the reader of the guide must also be a time traveler. In fact, it is not unlikely that this guide has been clandestinely placed in your path precisely to create a time traveler out of you—probably by a You From the Future, in an attempt to shape your destiny.

That's where QUAN+UM comes in. Trust us:

YOU SHOULD NOT READ THIS BOOK. STOP IT. NOW.

Here at the Qualified Users and Negotiators of Time Travel Universal Ministry, we must fiercely discourage any further interest in time travel. For one, the authors are enjoying the ride quite enough without the bridge and tunnel crowd gumming up the works with long lines and an endless barrage of flashbulbs (history is not your personal museum tour). And with the exception of one particularly annoying intern, most of us here at QUAN+UM have dedicated our lives to the study and protection of the space-time continuum and the support of officially sanctioned time travelers. You may think that riding a triceratops is hilarious—we have respect for that triceratops. Without us, future generations of time travelers would have to find out the hard way why triceratops-riding is for work and study purposes only.

Plus, let's face it: You shouldn't be time traveling. You don't even have any credentials. The long and short of it is, you're probably going to screw this up. And in travels through time, unlike regular old boring life, a mistake can mean a lot more than an unsightly mustard stain. Ever imploded a universe before? Well, now is not the time to start.[1]

Furthermore, if you are that rare exception—that one out of ten—it means that you are not a time traveler. Which means it is not too late to stop you.

---

1 Universe implosion merely theoretical. As of this writing, the universe has not imploded due to time travel.

The authors, despite failed personal goals and dwindling wages, will at least succeed in stopping one more fool from taking a chance to screw up history, kill him- or herself in horrific fashion, or disallow the evolution of puppies.

That's right: If you read on and unlock the secrets of time travel, you could be responsible for the dissolution of every puppy that ever exists—past, present and future. And we'll tell everyone it was you.

. . .

Okay, well, what if we told you that we would kindly like you to not read this guide, because if you do, and it really has traveled through time before, that's technically stealing. Maybe even from yourself, but especially from us. Instead of each new time traveler buying his or her own book, you just recycle the same beat-up copy over and over again throughout time. Time travelers have been known to hide copies in medieval castles to one day be excavated, or loan them to the Time Lords in the distant future. We give you time travel, but you can't pay for a simple guidebook? That just doesn't seem very fair.

We're giving you one last chance:

PUT DOWN THIS GUIDE, STEP AWAY FROM CERTAIN FOLLY AND ABANDON ALL NOTIONS OF SPACE BATTLES, GAMBLING RICHES AND BEING WORSHIPPED BY CAVEPEOPLE. THEY MAY SOUND REALLY AWESOME—THEY ARE NOT. WE MEAN IT.

. . .

. . .

. . . And you're still here.

Well done, sir/madam/sentient robot/super-intelligent ape/ other: You have passed the first psychological test! This was a necessary assessment of your fortitude, because as a time traveler, you're about to see some seriously messed-up stuff. Truth be told, your reckless abandon and disregard for mistakes are the stuff time travelers are made of. We need someone like you to measure the interior gravity of black holes, work out which parallel universes have atmospheres made out of lava, and test alternative time machine fuels for mutagenic biological properties.

Within these pages, you will find all that you need to successfully navigate time. The first half of the book is your primer, a textbook to the science of and necessary information related to creating and sustaining time travel for your safety. You'll get briefings on various time machines, the methods of time travel best known to humanity, and a great many samplings of the idiots of time travel's history, whose failures have helped to make your death less immediate.

The second half of the guide is your field survival manual for various periods of Earth's history, where you will find everything you need to know to blend in, communicate, fight, run away and repair your time machine in any era. Study up, or use it for reference on the fly. Note that this technique requires that you've first mastered the art of reading while sprinting and flailing your arms in horror.

Welcome, intrepid traveler, to *So You Created a Wormhole: The Time Traveler's Guide to Time Travel.* Just, please . . . make sure you paid for it.

1

WELCOME TO THE SAUCY MULTILAYERED BURRITO THAT IS TIME TRAVEL

CONGRATULATIONS! YOU'RE DOING it, reader. You're reading this guide and are thusly entering an incredible new world of temporal displacement—where danger, excitement and era-appropriate body hair lurk around every corner.

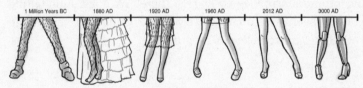

| 1 Million Years BC | 1880 AD | 1920 AD | 1960 AD | 2012 AD | 3000 AD |

Women's Leg Hair Throughout Time

Like those before you, the reasons driving you to take this brave step into the unknown are likely as myriad as they are idiotic; the path before you, as unnecessary as it is inadvisable. Welcome to time travel!

But just what is the fourth-dimensional road ahead? How does it work? Will you require a towel or just a change of socks? Should your time machine include a working toilet?

These and many more questions lie before you. Without proper knowledge, you'll be doomed to repeat the mistakes of the many, many, many, many[2] travelers before you who have met incredibly horrific, often totally unexpected, and sometimes technically impossible ends.

But you have what they don't. You have this guide.

And because of that, you're about to have a fundamental understanding of what time travel is. And that is what time travel is, in fact—the ability to travel. In time.

And not just in time, but through time as well as on time. Time factors so greatly into time travel that it's in the name. If there's one thing you should know about time travel, it's that time plays a key role.

And time, as it were, is on your side. That is to say, when you're traveling through time, it's like you're beside it. While the rest of us are caught in a river of time, being swept downstream in one direction, you stand boldly on the riverbank, moving up- and downstream at will, throwing yourself in and out the current as you choose.[3]

Anyway, that's time travel. Travel through time. You can go to the past, like two days ago, and you can go forward, like the day after tomorrow. But why stop there? We don't mind telling you that such an unimaginative venture would rank you among the worst time travelers on record.

TIME WHAT NOW?

Just what is time? This is a delicate question. It has been described in many ways: The Fourth Dimension. Half past twelve. A human construction maintained to ensure trains arrive on

---

2 *Many.*

3 Or often as fate dictates. As it turns out, you may have no say. (See section in Chapter 4 "Paradox by Predestiny.")

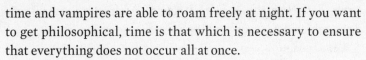

time and vampires are able to roam freely at night. If you want to get philosophical, time is that which is necessary to ensure that everything does not occur all at once.

However we describe it, to be a time traveler it is key to at least understand the ideas of past, present and future.

As Einstein explained in his Theory of Special Relativity, we all have our own personal time. You're always in the present, even if you're in the past or the future. In other words, you can never be in your own past, even when you're in your own past.[4]

How you experience time will never deviate much from how you've always experienced it. So go ahead and erase "fast-forward through hangover" and "last longer in bed" from your time travel bucket list. This is time travel, not advanced-level druid sorcery.

Another wrinkle: Time travel does not mean do-over. When you travel back, you won't be younger. When you travel forward, you won't be wiser. Want to finally make good on your promise to "totally bone" Mrs. Cook, eighth-grade math teacher? You can! But not as an eighth-grader. The best you can do is to bag her as Now You, or stand idly by as Thirteen-Year-Old You remains trapped after school for unnecessary tutoring because of the NARB[5] anchoring him to his desk.

But because you're as sharp-witted as you are well dressed, you're probably saying: "Hey! But what if I travel back and change the past? Then the future will be different! . . . Right?" A coy way of suggesting you might run your hands through Mrs. Cook's perm after all and, sadly, a rookie faux pas: The events following your alteration to the past would indeed be different, but you would not be there to live out your own life in this new, divergent manner. Perhaps you travel back to the future to see the results, or stick around to watch everything unfold—but you can never—

---

4   Footnotes are a good example of this. Are you at the end of the previous paragraph, or at the start of the next? Neither. You're right here, reading this footnote.

5   No Apparent Reason Boner.

quite—be—there. Unless you kill your past self and take your own place, which don't even act like we just suggested to you.[6]

Let's recap:

- Time perception remains constant, no matter how much you hop around through time.
- Reliving key or mundane life moments is not possible unless one kills oneself.[7]

Now that we have a fundamental understanding of how time works, let us journey forward and find out just how we got to this awesome benchmark in human achievement.[8]

A BRIEF(ISH) HISTORY OF TIME (TRAVEL)

Though theories abound as to how time travel works and just how awesome it is and has the potential to be, the truth is that science is way behind practice in this regard. Most of time travel remains unproven. Technically, you aren't even supposed to be able to do it. But tell that to time travel heroes such as H. G. Wells, John Connor, the Langdons,[9] actor Paul Walker, Timecop Jean-Claude Van Damme, the Teenage Mutant Ninja Turtles and the creators of Superman. And that's to say nothing of the unsung and unknown thousands who have thrown themselves into the gaping jaws of time, only to be torn to shreds by the razor teeth of failure.[10]

---

6 Revisionist histories get a bit sticky in Time Travel Methods in Chapter 2. Also more on self-killing in Chapters 4 and 5. Quick summary: NOT GOOD.

7 Time Travel Rule Number One: Interacting with your past self is inadvisable, as it can cause psychological damage, as well as awkwardness at parties, *Parent Trap* hijinks and the dissolution of the Universe.

8 Specifically, we mean this guide. But also time travel.

9 A family known for their spectacular failures in the field of time travel, widely discussed at www.thetimetravelguide.com.

10 This is a metaphor. It can also really happen.

The fact is—we can time travel, we do time travel, and we will continue to time travel, even if Science is a Negative Nancy about time travel. The remainder of the chapter will be spent with the two figures at the pinnacle of the Time Travel Mount Olympus[11] and their discoveries: Albert Einstein and Emmett "Doc" Brown.

ALBERT EINSTEIN AND HIS RELATIVE SUCCESS

There was a time when shooting about the quantusphere in an airbag-free steel cage was only the stuff of imagination. In H. G. Wells's *The Time Machine*, his "Machine" was little more than a souped-up stationary bike without so much as a nuclear power source or a protective outer casing, which—as any early adopter can tell you—burns the skin a great deal. Melts it right off. Bone, too. In fact, no early adopter could tell you this, because they're all dead. Quite embarrassing for Mr. Wells as viewed through the lens of retrospect, although we're starting to think he made the whole thing up, the bastard. Mark Twain, on the other hand, got us a bit closer to the actual science by sending the main character of *A Connecticut Yankee in King Arthur's Court* to the past via a concussion, which is much easier and not quite as painful.

It was not long after when a German-born man living in Switzerland at the turn of the twentieth century had a revelation that would become the stuff of slacker legend. A failure at mathematics in the realm of academia, Albert Einstein took a job as a patent clerk to allow himself as much free head space as possible to ponder a conundrum he'd found some years earlier in a children's book. Yes, a children's book. One with a high pretty-pictures-to-words ratio. Anthropologists and stoner recent-grads for centuries to come would ponder how, exactly, with limited technological resources, Einstein managed to explain this life choice to his parents.

The question proposed in the children's tale? "What would it

---

11 Located somewhere in the Rhineland. Or California. There's some debate.

be like to outrace a telegraph? To go faster than electricity?" Before Einstein's name became synonymous with "genius," and quite before that damned poster of him sticking his tongue out was Scotch-taped to the walls of nearly every high school classroom in America, Einstein pondered this vaguely imaginative problem for nearly ten years. Have you, faithful reader, ever thought about anything for ten minutes, much less years, aside from the proper way to install a roll of toiler paper?[12] Ten years of life in shitty apartments, re-wearing dirty socks, eating powdered cheese Easy Mac, enduring a lackluster sex life and taking public transit to work—he was either destined for greatness or destined for failure, a razor-thin tightrope walked by nearly every theoretical physicist thereafter. He was, gen-Xers would later say, "like, way ahead of his time."[13]

But it was, so the story goes, on one of those many trips through the city on a urine-scented bus that Einstein finally unleashed unto the world the storm of brilliance within his mind (and, consequently, relieved his parents a great deal about his mounting student loan debt).

You see, Einstein's bus was making its way past a clock tower in the center of town.[14] It was then and there, the smell of Swiss cheese and bratwurst on the air competing with the bus's thick hobo-musk, that Einstein wondered what this ticking clock would look like if his bus suddenly shot off at the speed of light. The light coming from the clock, he reasoned, would never reach his eyes, always trailing behind him at the same rate. It would appear to him to be still. His pocket watch, however, could he

---

12   "Over" is the correct answer, as proven by quantum physics in 2049.

13   Not a pun.

14   Nearly every influential time scientist has, at one point or another, forged a very special relationship with, or had an inexplicable affinity toward, an oversized clock of some kind. It should stand to reason that if you plan on making a dent in the tin can of quantum physics, consider either moving to a town with one or erecting your own (we mean a clock tower—the tin can was figurative).

have afforded one,[15] would tick along as usual because it would be moving at the same rate as our dear friend Al.

And with that, Einstein was on his way to the Special Theory of Relativity. Had he been more daring, Einstein would have simply started out by slapping rockets to said bus or rubbing his socks against the carpet to generate a massive static-electric charge, which was later found to be a means of generating temporal dissonance.[16] As it was, he quit his job,[17] got published, won the Nobel Prize and set about the first post-Newtonian paradigm shift in physics. It would have to do.

While the list of Einsteinian anecdotes and accomplishments is longer than all of his unfurled, pube-like hairs tied together end-to-end (for example, he married his first cousin), the focus here will be his three largest contributions to theoretical physics as they apply to time travel:

1. The Special Theory of Relativity
2. The General Theory of—

Actually, let's hold there for a minute. Did you catch that? Guy married his cousin. This teaches us one of two lessons: Either Einstein knew something we didn't and you should take a closer look the next time you're at the beach with your extended family, OR, if you are married to or plan on being married to your own flesh and blood, at least be a genius or an awesome athlete, as society will be more forgiving of just how repulsive you are. Anyway. Einstein's three largest contributions to theoretical physics as they apply to time travel:

---

15 He couldn't. Just go with it.

16 Time traveler reports indicate that he did indeed try this, but his hole-riddled socks, which he was too guilty to ask his wife to repair (we told you—guy's sex life was lacking) left gaps in the circuit that prevented the accidental ripping of spacetime. Thanks, Einstein, for setting the world back a hundred years.

17 Don't quit your job.

1. The Theory of Special Relativity
2. The General Theory of Relativity
3. Wormholes

The Special Theory of Relativity

The easiest way to explain Special Relativity is to use what is known as the Damned Dirty Ape Disillusionment Postulate. In this scenario, everyone on Earth is moving at roughly the same speed. Some travel by bullet train to work and others sit in their parents' dimly lit basement reading improbable guidebooks—but for our purposes, everyone is going slowly enough (the speed of Earth hurtling through space, more or less) that their speeds can be considered like. We'll call this speed: 17. Now, say you're an astronaut, and say you're on a super-important space mission in space, and say somehow you get separated from the space station and also somehow rocket off through space at nearly light speed. We'll call this speed: c-1.[18]

You travel at c-1 for ten years. A fair chunk of time as it relates to the human life cycle and an uncomfortably long time to pee through a tube. When your ship finally runs out of spacedust,[19] you slow down and see a marvelous planet before you. You crash-land (because you forgot how to pilot the damn thing) and find a world eerily like Earth, but with unfamiliar landscapes and a populace made up entirely of ape people. After being captured, escaping capture and possibly engaging in sexual relations with monkey natives, you find yourself on the beach in your skivvies. You come across what appears to be the Statue of Liberty half-buried in sand and it finally dawns on you: The planet you're on is Earth.

How could this have happened? You were only gone for ten years. Was it the Russians? Quite possibly it was the Russians. But the real answer is Special Relativity.

---

18 That's *c minus one*. What, you're thrown off by a letter? Did you even finish algebra? C stands for the speed of light, which, for reasons we won't get to, nothing can go faster than due to $E=mc^2$. You'll never see us mention this equation again, so stop being so proud of yourself for your vague recognition.

19 This is powering your spaceship.

Phil Hornshaw & Nick Hurwitch

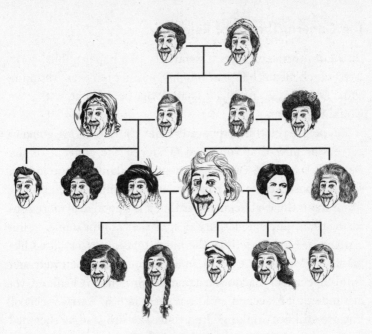

The Einstein Family Tree

See, while you were in that spaceship, hurtling through the cosmos at an alarming clip, the rest of Earth carried on at speed 17. The faster you go, the slower time progresses for you relative to those at slower speeds. So although only ten years passed for you, a good deal more time passed for the Earthlings—since you were going near the speed of light, your time was extremely slow as compared to theirs. Meanwhile, the people of your home planet were overrun by hyper-evolved apes, a staggering blow to both Darwinists and Creationists in equal measure. That's Special Relativity: the easiest and most common method of time travel and Einstein's first major contribution to temporal displacement.[20] The downside? It's one-way: future only.

---

20 BONUS TIP: Remember the bullet train? If the guy on that bullet train never gets off, he's going to live at least a few days longer than the guy in his basement. That's right! Special Relativity can help you *outlive* lazy people. Try it.

The General Theory of Relativity

Einstein proposed General Relativity in a paper published in 1915, which means he spent roughly another ten years thinking about another new problem. That's twice he did that, so that we wouldn't have to.

But before you think, relieved reader, that things are going to get easier now that you have mastered the concept of the Damned Dirty Ape Disillusionment Postulate and the idea of relative, observer-based time, know this: The least confusing thing about General Relativity is that it is a good deal more confusing than Special Relativity, despite the calming name.

Even in Einstein's time, the name mix-up underscored humanity's need for time travel. Imagine you had named your first child "Special." What do you name your broodingly brilliant, yet misunderstood, second child ten years later? "Extra-special"? You are still not thinking like a time traveler! Had he then had the capability, Einstein would have done as any rational time scientist and gone back in time, while his past self was sleeping or out bobcatting,[21] and changed the name of "Special Relativity" on his original paper to "General Relativity" and then traveled back to the future and named "General Relativity" "Special Relativity." As you can see, this solution would have been a good deal less confusing.

As it stands, the mint chocolate chip of modern physics has persevered with a name like vanilla. Einstein's original aim for general relativity was to account for Newton's Laws of Motion in relativity's framework. For Newton, gravity is constant, and the frame of reference is irrelevant. But Special Relativity teaches us that gravity, frame of reference and other external factors are essential to our understanding of physics and the manner in which the universe operates. In accounting for these factors, gravity primary among them, General Relativity would for the next century

---

21 Hunting for your own hot cousins at the bar.

Phil Hornshaw & Nick Hurwitch

lead to countless equations named after whoever came up with them and quite a few startling conclusions. We'll spare you the equations and their names (which you'll be responsible for on the test)[22] and jump right into what this all means.[23]

The General Theory of Relativity finds that acceleration and gravity are indistinguishable. That is to say, the force of gravity merely acts as a form of acceleration. Theoretically, that means gravity can aid in the forward-moving form of time travel described by Special Relativity. But it would take a massive amount of gravity over an extended period to have any real effect. They pull this off in *Star Trek IV: The Voyage Home* and actually send themselves back in time by using what they call "the slingshot method." They . . . well, slingshot themselves around a star in the starship *Enterprise* and use the curvature of space to—you know what? This is bullshit, so don't try it. *Star Trek IV* is just a movie.

Taken to its extremes, General Relativity also postulates the existence of black holes and singularities—which are, in fact, quite real. This is significant because being sucked through a black hole or into a singularity could transport you through time and space. More often than not, you will simply be crushed or instantaneously vanquished from existence as your atoms disperse and become one with said gravitational phenomenon. But in the event that you survive the effect of a thousand suns' worth of mass crushing down upon your fragile bone structure, you could find yourself in a mysterious new time and space.[24]

The last spicy tidbit that General Relativity offers mankind is one of economy. You may have noticed that in the preceding paragraph we used the phrase "time and space" not once, but twice. What a waste. Einstein found out for us that time and space are in fact one. Spacetime, he called it, and so do we. This

---

22 There's a test.

23 If you really are interested in time travel, you might as well get used to this approach. Jumping right in is how we do things.

24 In Chapter 2 we'll get into why this is probably bad.

has a couple of consequences beyond that of the grammatical. Consequence The First is that matter affects the space around it. If you imagine space as your dirty bedsheets, and a planet or star as the pile of naughty magazines you unintentionally left on your bed for your mom to find, you'll see what we mean. See how the bedsheets curve? How the presence of the enormous stack of pornography warps the space around it? This is also observable in space. For example, it takes a beam of light longer to pass by a star or planet than it does empty space. But because the speed of light is constant (meaning it doesn't slow down, not for nobody), we know that the light is only slower because it has to travel along a curved region of space.[25] The light beam has farther to go.

Consequence The Second is so scientific, we can put it in an "If, Then" statement: If space and time are one, then we might very well be able to travel through time by traveling through space. Lady[26] and Gentlemen, we give you . . . wormholes.

Wormholes

If you read the title of this book, you may have gleaned that wormholes are an important facet of time travel. Whereas Special Relativity time travel is fraught with uncertainty and an inability to return home due to its ape-future one-way nature, and General Relativity is wildly ineffective and needlessly complex, wormholes are a surefire way to shoot from one place to the next, and one time to another, all from the safety of your time machine.

In essence, a wormhole is a subatomic hole connecting one infinitesimal point to another distant and seemingly unrelated

---

25 If this is difficult to conceive, think again of your dirty bedsheets and porno mags.

26 Not a typo—there really is only one female reader of this incredibly nerdy book.

infinitesimal point elsewhere in the universe. The easiest way to demonstrate this is with the following exercise:

•   •

The two points above are nowhere near one another. Clearly. But if you were to curve the paper, and fold the page in half so that the points touched, you would in effect be creating a "short-cut" from one point to the next. Wait—you didn't actually fold the page in half, did you? That was a hypothetical! At no point did we directly instruct you to fold anything. Now the book is damaged, its value in future estate auctions is severely diminished and there is no chance you can return it to the bookstore for the refund you're probably starting to believe you deserve. Well done.

Nonetheless, the point is made. Wormholes exist all around us, and time traveling with some measure of control now becomes a simple matter of stretching an incomprehensibly small space into something large enough that you can walk through before it disappears with you trapped inside, or partially inside and you lose an arm or something.

It would be many decades after Einstein introduced the idea of the wormhole before it would be practically applied to time travel. He would die in 1955 in Princeton, New Jersey, at the age of seventy-six, long outlasting his wife Elsa, who died some twenty years earlier. She was, for the record, also his cousin.

EMMETT BROWN AND THE HIGH-VELOCITY DELOREAN WORMHOLE TRAVERSAL DEVICE

By 1955, nobody had figured out how to build a vehicle that could travel as fast as a beam of light—although a few military test pilots had broken speed records zipping about New Mexico in crashed flying saucers retrofitted for human use. They were at the very least scaring the crap out of people.

Meanwhile, the scientists of the day found themselves side-tracked. Einstein's super-speed requirement for time travel seemed to many to be impossible to achieve, and they therefore turned their attention to unleashing the unholy wrath of the subatomic and using it to blow things up.

It was not until Dr. Emmett Brown, a rich eccentric from Hill Valley, California, came along that the next significant stride in the creation of an actual working time machine would occur. He was the inventor of a device known as the flux capacitor, an electricity-based quantum vortex generator. Using an astounding amount of power—1.21 gigawatts, a surge so enormous it can be achieved only by nuclear fission or a bolt of lightning—Brown found he could momentarily bend spacetime enough to create a wormhole from one time to another. With accuracy.

Eventually, it would lead to a stylish time machine, a rash of sequels and a whole lot of hijinks.

But Brown wasn't always a successful, jet-setting wormhole traveler. Before he was Doc Brown, Time Machine Inventor, he was Emmett von Braun, Don't Talk to That Crazy Old Man, Kids.

Emmett von Braun, as he was once known, was a relative of German rocket scientist Wernher von Braun, who was brought from the Fatherland shortly after World War I and contributed greatly to the military's efforts to affix rockets to nearly everything. Unable to step out of his father's shadow even late in life, Emmett eventually changed his name to "Brown" and began referring to himself as "Doctor," though he had no higher education credentials to speak of.

The flux capacitor and its resulting time machine would be Brown's legacy and life's work, but before its perfection in 1985, he was in fact a dismal failure as an inventor. Among his various nonworking inventions: a mind-reading helmet purported to inflict significant brain damage in certain test subjects; a speaker amplification system that once exploded during an easy listening folk festival in northwest Pennsylvania, scattering concertgoers into several adjacent counties; and an inexpensive water purification system that later was found to, at certain temperatures, accidentally restructure water molecules into something akin to citronella oil.

Brown wasn't even necessarily interested in time travel until he accidentally gave himself a concussion in a one-man household accident. According to his unofficial biography, *Emmett Brown: Unlocking Genius Through Accidentally Falling Down* by Erin Williker, Brown was inspired to create the capacitor after attempting to hang a wall clock in his bathroom. He stood on his toilet in order to reach sufficient height, but lost his balance and toppled backward, cracking his skull against the sink.

The injury wasn't serious, but Brown blacked out from the impact. When he came to some hours later, the image of the flux capacitor was floating in his mind. He dedicated the next thirty years and an obscene amount of his family fortune to making real what some neurologists might refer to as a "brain hemorrhage backfire."

THE VEHICLE

The flux capacitor harnessed energy and channeled it to create the mechanism of time travel—a functional wormhole stable enough to transport a person or dog through time. But as Brown soon discovered, the portal only remained open for a split second before abruptly slamming shut, bitch-slapping anything inside. The capacitor could indeed open the portal, but without enough speed to make it through completely, any object trying to traverse it would be disfigured in a manner not unlike the effects of jamming your finger into a cigar cutter or blender.[27]

After an Edison-like series of shortfalls, Brown found the De-Lorean (a car with a stainless steel body designed by an eccentric automaker working out of his own garage) to be the most

---

27 Your finger being the time machine. The other part of your finger being the part of the time machine that has disappeared into another dimension.

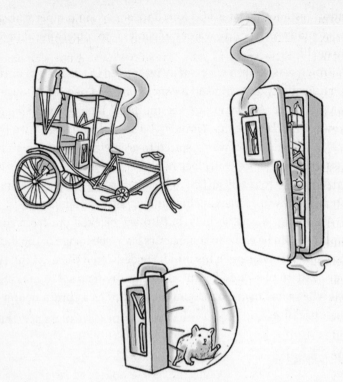

Failed Time Travel Traversers

effective vehicle for time travel. Its benefits were threefold: The steel body could better conduct the flux capacitor's energy to open wormholes around the car, and could also protect the occupant from the negative effects of a near-inconceivable electric charge. The third benefit was that it looked fucking rad.

All that remained was a matter of speed. Just how fast would the time machine need to go, and just how many lab animals would need to be slaughtered in the name of Science before the first successful wormhole jump in the history of mankind?

This is when Brown had perhaps his biggest discovery not involving a head injury.

Using some calculations that we're legally barred from sharing with you, Brown found the ideal speed for traversing the wormhole without being cut in half or blended by spacetime at large. One did, indeed, need to go fast to travel through time, but not nearly at the speed of light. The solution came not a lab hamster too soon.

The speed required: eighty-eight miles per hour. Any faster, and the machine would miss the portal entirely, causing the car to crash into a nearby shopping mall or town hall. Slower, and the portal would snap shut on the time machine, causing embarrassing, though instantaneous, death.

It was time to act. Brown struck a deal with a Libyan terrorist cell operating in California, agreeing to build an active nuclear warhead with the cell's modest supply of weapons-grade plutonium. He used the plutonium and the 50 percent up-front payment to make a working model of the flux capacitor, nearly thirty years after its original conception. Unfortunately, he would only have one chance to test it, the theory and the DeLorean time machine. This would be Dr. Emmett Brown's last stand against, and in the name of, Science.

After testing the device on his best friend, Einstein the Dog,[28] Doc knew he was on to something. He upgraded from animal testing to human teenager testing, and the rest, as they say, can now be purchased in a special edition box set at your local Wal-Mart.

Use of the DeLorean nearly caused irreparable damage to our timeline on at least three occasions, but Brown's trial-by-fire approach not only led to the first known successful manned time travels, but blazed a trail for every haphazard imbecile with a DeLorean and a death wish to stumble boldly after.

Doc Brown retired to a quiet frontier home with his wife, Clara, in an undisclosed location in 1888, but his work was not forgotten. It is his machine and Einstein The Guy's theories that

---

28 Little-Known Fact: Doc Brown and Einstein the Dog actually had a tense, slightly murderous relationship, culminating in Brown's reckless use of Einstein the Dog as a living test subject.

have led us forward into the era of time travel. And also backward. That's kind of the point.

In Chapter 2, we will take a closer look at various means of time travel and make a resounding endorsement for machine-based travel.[29]

---

29 If you've already attempted time travel based on the introductory knowledge contained *prior to the end of the first chapter*, this footnote waives our liability.

2

TIME TRAVEL MODELS AND THEIR RICH HUSBANDS

S O YOU'RE CONVINCED temporal relocation is for you. You have a base understanding of time travel, have familiarized yourself with its forefathers and are finally considering asking out your cousin. All told, you believe that you have taken a rather large bite from the saucy, multilayered burrito that is time travel. Well, we have some bad news for you: You got that burrito's sauce all over your chin. And by sauce all over your chin, we mean brains leaking out your nose—because your mind's about to get blown.[30]

You may have a handle on the flow of time and the general concept of moving in, on, through, around, next to and perpendicular to time, but here's a news flash: There's more than one way to end up in yesteryear. Or tomorrowtown, as the case may be.

The time travel guide–impaired are well known for lumping

---

30 Out of your skull. This is an analogy. Unless, of course, your brains really do come out of your skull while attempting to time travel. In which case, we told you so.

all instances of time travel together into some singular, magical means of getting from one remarkably different time to the next. But the reality is that the means of George Taylor's venture into the future to do battle with a race of super-intelligent apes was in fact quite different than actor Paul Walker's journey to Medieval France to prevent an enjoyable film adaptation of one of Michael Crichton's best novels from ever being made.[31]

Just as standard methods of travel carry different risks—running, for example, may cause you to trip and fall, and skipping may cause random passersby to punch your face—not all modes of time travel are created equal.[32] In fact, some are quite dangerous. Well, to speak plainly—they're all pretty dangerous. In the grand scheme of danger, time travel is one of the most dangerous things you can do, right up there between chuteless base-jumping and scorning a woman. Statistically speaking, only one in every four attempts at time travel does not result in death, and only one in six does not feature permanent maiming of some kind.

Some models of time travel, however, are considerably more dangerous (resulting in more burning or limb and skin loss) than others. The following means by which one might travel through time are organized by stability, or inversely as measured by the likelihood of death resulting from use:[33]

- Wormholes
- Relativity
- Quantum Foam
- Black Holes

---

31 See *Timeline*, Paramount Studios, 2003. Actually, don't. Read the book instead.

32 Turns out, the biggest risk to time travelers is time travel, and not time-traveling robots, as previously speculated.

33 This list by no means guarantees the success of any of the models. QUAN+UM retains no legal responsibility for any destruction of property, sanity, or molecular stability.

Phil Hornshaw & Nick Hurwitch

| Swimsuit Wormhole Model | Fashion Catalogue Relativity Model | Quantum Foam Hand-model | Blackhole Quasimodel |

The Four Models of Time Travel

Therefore, you can expect wormhole travel—the first model on our list and in our hearts—to be least likely to kill you. Conversely, the last model (black hole travel) is among the worst ideas you could ever have.

But this isn't all about your death, the vessel of your death and the gruesome and/or public nature of your death. You are reading this guide, presumably, because you intend not to die! At least not while time traveling. An important first step toward this theoretical non-death is recognizing what method of time travel is being employed and understanding its strengths, weaknesses and risks.

Familiarizing yourself with the varying methods of time travel will also help you determine how—or if—you can return home. Or, at the very least, diminish the probability of arriving in a distant time without some crucial ingredient, such as an ear.

WORMHOLES

Einstein's coolest idea since totally safe, limitless atomic energy,[34] the wormhole is a time traveler's best friend.[35] Almost any time one willingly travels to the past, a wormhole is the means employed. In fact, many other seemingly distinct methods of time travel actually rely on some form of wormhole to be possible.

If you paid attention in Chapter 1, you will recall what a wormhole is: a bending in spacetime that allows for much-abridged travel from one point to another. You will also recall what is required to create one: some form of flux capacitor or equivalent space-bending device, a whole lot of energy and the means to hold the portal open (like a Stargate[36]) or get through it before it closes (like a DeLorean).

There are a few factors that make wormholes your safest, sometimes cheapest and often most visually stimulating[37] method of time travel:

PROS

Hard to open, harder to get through

A wormhole is a hole in space and time and is therefore not actually something you want to be inside. The degree of difficulty involved in entry (and passage) means that you are unlikely to one day time travel through a wormhole, or accidentally be blipped into oblivion, without first giving it some serious consideration.

---

34 Unfortunately, this turned out to not be an actual, possible thing. But still a very cool idea.

35 Unless you manage to befriend and subsequently affix a saddle to a dinosaur. See section "SURVIVAL GUIDE: PREHISTORY: DINOSAURS, Riding."

36 Stargate (n): 1. Machine for generating a wormhole between two points in space, but not time. 2. So-so movie starring Kurt Russell. 3. Far better TV show starring MacGyver.

37 Read: easiest to pitch when the time inevitably comes to sell the movie rights.

Phil Hornshaw & Nick Hurwitch

RELATED: Require special materials to create

The plus side of needing something really expensive and scary to make your machine work—like, say, plutonium—is that you have to be at least smart enough to (1) obtain plutonium (without dying) and (2) build a machine that uses plutonium (without dying).[38] This drastically reduces the number of idiots time traveling by wormhole, present company notwithstanding.

Instantaneous time travel, backward and forward

By their definition, wormholes allow travelers to move from one point in spacetime to another—and the less time spent traveling, the lesser the likelihood of an accident. Consider a standard mode of transportation, like driving: The longer you're in the car, the greater your risk of crashing.[39] Not to mention, the usurping of time is more or less the whole point of time travel.

Backward travel possible

In the wonderful world of wormholes, traveling back in time is indistinguishable from traveling forward.

Created by several means

Though it's more fun in ways we won't here enumerate, you don't necessarily need plutonium to create a wormhole—anything that generates a whole lot of energy, like lightning or mass quantities of feces-eating bacteria,[40] can do the job in a pinch. And although 1.21 gigawatts of energy is nearly impossible to generate, nearly impossible is not impossible. Time to pull up your trousers and start thinking like a time traveler!

---

38 Deaths incurred during the acquisition of radioactive materials and construction of machines that employ them are not counted as time travel deaths by the QUAN+UM Time Travel Death Comprehensive Statistical Survey Bureau. Neither is cancer.

39 Additionally, nearly 85 percent of all accidents occur within one mile of the home. We suggest you time travel just outside this radius.

40 See section "SURVIVAL GUIDE: PREHISTORY: TIME MACHINE REPAIR, Building Your Unfossilized Coprolite Battery."

Requires a great deal of energy

Time travel can become an expensive endeavor. Plutonium doesn't grow on trees on most habitable planets and you can't exactly pick up a mad scientist down at the local pound. And while when you consider the fact that you can build a wormhole traversal module out of something as simple as a car, wormhole travel seems relatively cost-effective . . . have you ever tried to purchase lightning? We have. It's not cheap. And they make you bring your own bag.[41] Also not cheap.

Measurable possibility of exploding, flaming death

Did we say possibility? Because the word we meant was probability.[42] Just because you're providing a practical application for bolts of lightning or unrefined plutonium doesn't mean it comes safe out of the box.[43]

Must navigate time and space

While it's one thing to say "I'd like to visit Egypt, 2543 B.C.E.," it's quite another to make sure you don't arrive on the inside of a sand dune, or worse: the emptiness of space. Though to us the presence of Earth feels constant, its cosmic space dance could leave you out in the cold just like so many disappointing middle school dances. Make sure your wormhole traverser accounts for the movements of the heavenly bodies.

RELATIVITY

Einstein's Theory of Relativity provides a fairly handy method of heading into the future, without the necessity of folding space.

---

41 Lightning-safe bags available in the QUAN+UM online store at http://www.thetimetravelguide.com on the World Wide Web.

42 Actually, we mean near-certainty, but the publisher was worried a guide to something nearly certain to cause death wouldn't sell.

43 Assuming you're lucky enough to have a box.

In fact, relativity operates on a much more conventional plane.

We got into relativity a bit in Chapter 1. Here's the gist: the faster you travel, the farther into the future you will wind up.

Einstein postulated that at just below the speed of light, time would slow down for those traveling in relation to those traveling at slower, normal speeds. There is a ton of math that shows how this works, but as a talking Barbie toy once said, "Math is hard."[44]

Here's what you need to know: If you build a spaceship capable of going nearly as fast as the speed of light, you can travel into the future. If you can go faster than light, you'll end up way in the future. But keep in mind that the speed of light is approximately 10^{23} times faster than that thirteen-minute mile you ran in the eighth grade—so, no, that wasn't technically time travel.

PROS

Easy to understand

Go fast, see future. It doesn't take a rocket scientist to get it.[45]

Figure out space travel and receive time travel as a free gift!

Faster-than-light travel is the Holy Grail of space travel.[46] While scientists in the twentieth and twenty-first centuries write off time travel as an absurdity, they continue to work diligently on space travel, inadvertently paving the way to the future via relativity by 2045, when the first FTL drives are developed. Put simply: In your face, Science. Score one for Science.

Finally, a use for mathematics

---

44 That's why you have all those Barbies today that are presidents, biologists, and space marines. Hard to believe that with the Apocalypse on the horizon, this is what concerned American citizens in the 1990s.

45 Doing it, however, does take a rocket scientist.

46 Quite literally. When it's eventually found in the attic of an Atlanta, Georgia, apartment, it's discovered that the Cup of Christ is inscribed with the equations necessary for FTL transit, the secrets to quantum computing, and a killer recipe for hummus, which adds some pep to the otherwise bland Body of Christ.

Unlike wormholes, which require a great deal of work and, in many cases, blind experimentation to discover where you are being sent, relativity is always the same. Einstein defined the equations necessary to determine how much time would slow down for the traveler in relationship to other objects moving through spacetime at slower speeds. Just run some well-known equations through a computer and you can determine exactly how long you need to keep your spaceship's FTL engines burning. Voilà, you're in the future. No getting chopped in half by closing wormholes, no intense gravity from black holes, no accidentally winding up in the middle of empty space, suffocating as you curse your own stupidity with your dying gasps.

CONS

You're never coming back

We indicated as much in Chapter 1, but if you employ relativity as your means of future-travel, you will not be booking a round-trip ticket. Best-case scenario, you'll arrive in a future in which wormhole-based, backward-travel-capable time machines are readily available for purchase or theft. To survive in your new permanent future-home, reference the survival guide at the back of this book.

Time-consuming

Unlike the other three methods of time travel, relativity travel does not transport you instantly through time. When you travel faster-than-light, you're going into the future basically because time for you is slower than time for the rest of Earth. You age a minute, they age two minutes. You ride around in your spaceship for ten years, and you arrive one hundred years later on Earth. Yeah, that's great—you went one hundred years into the future. It only took ten years to do it. Which, depending on how impatient you are, could have been a pretty colossal waste of time. It's almost as if you need some kind of secondary system of technology to help deal with the slow passage of time . . . something . . .

Like cryogenics!

Cryogenics is technology that works to suspend animation in its subjects—which means, in essence, freezing yourself. Travelers on long space voyages can climb into big freezing tubes, go to sleep and wake up ten, twenty, even thousands of years in the future, having experienced nothing more than a hearty nap.

Cryogenics is a pretty great idea on paper; when it works, it works like magic.[47] But the reality is more like attempting to powernap in a public park and not expecting to wake up at knifepoint with your laptop bag missing.[48] Even though you get to sleep your trip away, wind up in the future and never bother with any of the technical complications of astronaut work, going mad in the repetitive depths of endless space, or running out of freeze-dried rations, you're basically trusting your life to a machine while you sleep. It has to keep you alive, frozen, fed and breathing. The list of things that can happen to a giant metal pod hurtling through space at light speed for thousands of years could fill another volume of this guide.[49]

FTL travel means really, really faraway travel

It only takes light about eight minutes to get from the Sun to

Cryogenically Frozen Bill Chumperdink dies quietly in his sleep.

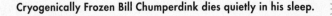

47 Excepting, of course, the grueling hangover and black tar lurching.

48 Which is probably for the best: It made you look like a sissy.

49 Which we are willing to sell you.

the Earth. Pretty fast, especially when one considers that the Sun is 93 million miles away. In the same distance you could circle the Earth 3,734 times. Run 3,576,923 marathons. Place several trillion boogers end-to-end.

That's less than eight minutes in your faster-than-light ship and suddenly you are inside the sun's molten core. Point being: When you go faster than light, you end up somewhere. And that somewhere is far, far away. So when you consider relativity flight, your two most careful considerations (aside from those relating to cryosleep) are:

1. A clear path
2. A definite destination

Much will change in the universe over the course of your journey. The extent to which such careful planning will help you arrive safely in the future is limited and theoretical at best. But you might as well try.

QUANTUM FOAM

The concept of quantum foam is a tricky one. And although here at QUAN+UM we are dedicated to the rigors of "science" and "accuracy," our editor routinely reminds us that our core reading audience is made up of simpletons from the past reading at a third-grade level.

We will go out on a limb here and say that when you read "Quantum Foam" in the header above, your brain thought very little of "Quantum," but upon reaching the second word immediately produced for you an image of bubbles. Soap suds. Squishy packaging material. The innards of a stuffed bear. And your brain's production of that image is precisely as theoretical physicists[50] intended.

---

50 Who are much smarter than you.

Phil Hornshaw & Nick Hurwitch

What the word "foam" is getting at is a physical "something" that can only be loosely defined. A construction of something tangible that, at its most fundamental level, is actually quite chaotic. Within the foam, no two bubbles are exactly alike; some bubbles pop, others form; and if you play with it enough, you can shape the foam to resemble something else entirely. This, believe it or not, is pretty much how the universe is built.

Let's call in an example: Though you can look at this book and see it in your hand, flip each page, graze your hand over its exorbitant price tag and slam the book down onto your desk in anger, on a subatomic level, all of these things hardly exist at all. Electrons are spinning, quarks are gyrating (suggestively) and even smaller particles are jiggling about inside each of those. And when we get down to a level that unfathomably small, particles don't even exist in a single, definable location. They have equal probability of being in one place or another, and sometimes are in both places at once.[51]

And so, believe it or not, quantum foam—or the state of the sub-sub-atomic just described—is a lot like foam: particles of different sizes unsure of where to be, zapping in and out of existence and looking not at all like something tangible, with a bunch of empty space in between. But somehow, if we have enough of this foam, and we back away far enough, we end up with tangible things—the universe as we humans experience it. Tangible things like time machines and burritos.

You may have forgotten by now, but somehow, believe it or not, this all relates to a method of time travel. In essence, it involves shrinking or being broken down atom by atom, shot through the infinitesimal spaces in the foam, then coming back out the other side in another time . . . and possibly another universe.

---

51 In fact, quantum physics is just a bunch of equations of probability—scientists look at really little particles and try to guess where they'll be and what they'll do. Think of the early twenty-first-century game show *Deal or No Deal*. But in this case, the "deal" is the stuff that makes up all matter and the "no deal" is another universe or something. People get degrees in this.

Isn't Quantum Foam Just Wormholes Again?

Before you get smart with us—a first, according to our records— and suggest that Quantum Foam sounds an awful lot like Wormholes, let us count the ways it is different. Two. There are two ways in which it is different:

1. Big vs. Small

Instead of stretching a wormhole out to shoot through it as normal-you size, with Foam you're being shrunk down or deatomized in order to fit through the infinitesimal hole within the foam. Boo-yow. Different.

2. The Principle of the Traveling Pants

Though we commonly refer to a wormhole as a portal, a better way to think about it may be that it is two ends— entrance and exit. What is "in between" is just a matter of getting from the former to the latter. Put simply: If you can open up a wormhole, you can theoretically take it to any exit, at any space, in any time. That said, if a wormhole is the spandex of time travel outerwear—comfortable, yet stretchy and sleek in nearly any size—Quantum Foam is the wood. As in wood pants. It's not very flexible. If it helps, you can still think about Foam travel as a portal with an entrance and exit—but bear in mind that that entrance and that exit are fixed. If you wish to reach a particular exit, you must take a particular entrance. Furthermore, there is reason to believe that you are not entering or exiting at all, but are instead using a gap in the Foam to reach another point in said Foam's existence, which on a quantum level exists at the same time as all other points in its existence, simultaneously. Boo-yow, redux. Double different.

Shaking your head side to side? You're headed in the right direction (that is to say, side to side). Please refer to the Pros and Cons below for further explanation.

PROS

As with a wormhole, can be done anywhere

Though the out hole is fixed, you are surrounded by in holes at all times. So, despite the two (2) key differences above, Quantum Foam travel does still allow for travel anywhere, at any time, provided you possess and can operate the requisite equipment and have an up-to-date, legible Foam map.[52]

Backward travel technically possible

In addition to forward travel. But only technically, because that all kind of depends. Quantum Foam travel takes the guesswork out of time travel by making everything a guess and none of it work. In other words, it decides for you. So maybe you're going to the past . . . then again . . .

That universe you're messing up might not even be yours

Photons of light interfere with one another all the time, even when they're not near one another. If you close off single photons, isolated away from all other photons, the result is the same interference pattern—as if all the photons were still there all along.

This could suggest that there are several universes beyond our own. If you send a person through Foam, they could wind up in another universe. In fact, there is so much Foam and are so

---

52 Which due to certain trade bylaws tend to only be printed in French Canadian. And, even if you happen to be French, Canadian, or both, they are notoriously difficult to read.

many possibilities, some scientists think there could be an infinite number of universes. Universi.

If there are infinite universes, it stands to reason that there are infinite universes that are wholly different from ours, but also infinite universes that are very similar, since you have to fill infinity somehow and universes where there are monsters or ape people or dinosaur cities are only, well . . . three ideas.

Therefore, when you time travel through Quantum Foam, you might not be traveling through time at all—you might be traveling between universes, some of which are in the past in relationship to your universe. Which means that there's an outside chance that if you make changes to the timeline there, your own timeline will not be affected: You're not operating in the past, you're just operating somewhere else entirely. It's the time travel equivalent of peeing on the seat in a public restroom, knowing you'll never have to deal with the poor angry sap who ends up sitting in it.

CONS

That universe you're in? Might not even be yours

While the potential of fudging up the present based on fudging up of the past is greatly reduced because you might not even be in the same universe anymore, you should also think of it this way: When you return "home," you may not really be back home. It could be one of the other infinite parallels that either (a) are slightly different or (b) already contain a you.

And if that's the case, you might not ever be able to get back to the universe where you started. It is roulette of the highest order and the lowest odds.

To make matters worse, in addition to being untraceable and slightly variant, these universes also carry a strong link to your original universe and to one another. This is commonly referred to as The Jerry O'Connell of *Sliders* Principle. Infinite though they may be, all the universes that make up the Universe are linked and make up one, all-things, all-times, super-ultra-mega Universe. So even though making mistakes in the past might not mess up the future, there is still a strong chance that it will.

Equally possible is that whichever of the parallels you end up in will look nearly identical to the one you left—so that pathetic life you left behind? Probably still there. Might even be worse. This is known as the Definition of Infinite Principle: If there are infinite universes, there are infinite universes recognizably different from your own . . . but also infinite universes nearly identical to your own, only differentiated by the size of a pencil, the girth of a bow tie, or some other minor detail or decision unrelated to you and your middling impact on the Universe at large.

You could end up a big pile of Pig Goo

This Con is a two-parter.

PART ONE: Deatomization-Reatomization process takes pinpoint precision. Quantum Foam–based machine operators typically incinerate and maim many hundreds of dolls and space apes before finally graduating to the homeless and crewmates. Taking someone apart who is moving, several universes away and often a moron—then re-creating them back at your machine—is a delicate task. It is not to be taken up by just anyone. Take QUAN+UM's advice: There is almost no time emergency worthy of allowing your cousin Merv to take a stab at operating the deatomizer.

Ask yourself: Would I allow Merv to perform emergency brain surgery on me right now? If the answer is **no**, he should not be de- or reatomizing anything. If the answer is **yes**, multiply that need by two, ask again, then—if the answer is still yes—at least make Merv skim the manual before proceeding.

PART TWO: There are existential crises inherent in deatomization-reatomization. When you lose an arm, you're still you. If you have your skeleton replaced by adamantium, you're still you, even if you forget your real name and start murdering people indiscriminately. If you have just about everything but your brain replaced by robotic parts, you're kinda still you, but are mostly a cyborg, like Robocop, a law enforcement professional the likes of which Detroit could have used several decades prior to his actual production. But if you are taken apart—atom by atom—and "reconstructed" . . . there is little evidence to prove that you are still "you," in the traditional sense. Where did all your original atoms go? Are the new atoms now "you," or are they borrowed limbs, like the body of Robocop at a more thorough, subatomic level? If you remember everything before your reconstruction, you might still be you, but the jury is still out. Just don't be surprised if around the water cooler you are soon known as "Reconstructed Limbs Boy," "De-Adamized," or "Hobocop."

Where the F*ck, Exactly, Are You Going?

Quantum Foam travel operates within specific linkages with the subatomic. Think of it like a tunnel: Once you build it, that's where the tunnel goes. You can go in and out either side as many times as you'd like, but they always lead to the same places. Thusly, the progression of traveling through Foam is generally as follows:

1. Where the f*ck am I going?
2. Where the f*ck am I?
3. How the f*ck do I get back?
4. Am I really back, or am I just "back" (f*ck)?
5. Bored today. Guess I'm going to Des Moines, Iowa, 2543 B.C.E. again.

So your frantic guessing and hurtling through spacetime like a spacetime dart toward a dartboard that might be made of lava or Grape Robots is followed quickly by either death, being trapped in another universe, or boredom.

BLACK HOLES

If you began making a list of things you would be willing to travel through—tunnels, forests, mega-malls, middle age, self-doubt—

chances are "collapsed star" would not appear very high on your list. But that is precisely what a black hole is: a collapsed star.

A star burns bright and hot because it is essentially a massive fusion reactor—burning through hydrogen, producing helium. A black hole results when a star fuses up heavier and heavier atoms until it collapses under the force of its own gravity. Everything that once made up the star becomes crushed so tightly, and the gravity becomes so remarkably intense, that not even light can escape. Light—the second fastest thing in the galaxy behind your relativity-based starship time machine. Some scientists believe that black holes might be a bit like wormholes, or even like the gaps in Quantum Foam: Because of the hole's gravity warping spacetime into a fold, flying into, and subsequently through, a black hole could mean you ending up in another space or time. Science fiction has speculated for years that finding a way to travel through a black hole could even help humanity access other dimensions.

Much more likely, however, is that a black hole is merely a really interesting way to annihilate yourself.

According to Einstein, a regular amount of gravity, say that of a planet, is enough to distort the space around it. The theory here is that black hole–level gravity must do all kinds of weird things to space and therefore time as well.[53] Going through one and successfully ending up in another time is not only unlikely, it is also undocumented and crushy.

PROS

Space travel again, with time travel side effect

Like the development of FTL drives, black hole travel is another example of time travel as an excellent space travel by-product. Go ahead and let other people build the exorbitantly expensive spaceships, antigravity and faster-than-light drives. When the technology is finished, time travel will only cost you whatever it

---

53 See section in Chapter 1 "Einstein and His Relative Success," which we thought you just read.

takes to get on one of these spaceships and hijack the bridge. Worst case, you are on a space cruise with the added possibility of infinitely painful molecular tearing.

Wormhole-like benefits

A black hole, if you can really fly through one, operates on all the same theoretical standpoints as a wormhole: distorted space, pinched-together fabric and instantaneous arrival. If you're in a pinch, can't scare up a wormhole and are not really a "frozen-sleep dice roll" kind of person, then a black hole becomes something of a viable option.

CONS

Spaghettification

Spaghettification, despite the joke your brain just made, has nothing to do with the effect of a spaghetti dinner on your bowels. It is in fact a real scientific term—need we remind you, this book is about Science and its authors are, in fact, time scientists. And it's not just any scientific term, but one coined by Stephen "Mother F——" Hawking, Universe-Class-A-Plus-First-Place Genius.

Spaghettification is an effect of the extreme gravity of black holes. MF Hawking says that when you near a black hole, the differences in gravity as you approach the center become so extreme over short distances that if you fell feet-first into the hole your head would experience one level of unfathomable gravitational force while your feet would get the full wrath of the universe's incredible sucking power. You would literally begin to spaghettify: Your feet would stretch out from your toes up and eventually the molecules would break apart and be pulled into a sort of wet noodle shape as you continued your plummet of doom.

The subtext here is that you don't want this to happen to you. And the between-the-lines of that subtext is that if it does, for the love of all that is holy, don't go feet first. But if you are going to go through a black hole anyway, you're going to need some kind of special vehicle that allows you to even attempt to navigate up to the black hole, much less travel through it.

Phil Hornshaw & Nick Hurwitch

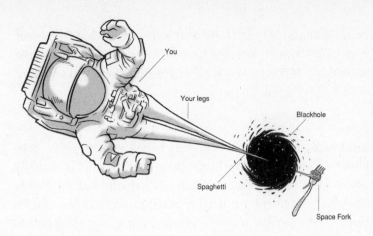

You

Your legs

Blackhole

Spaghetti

Space Fork

Need for a gravity drive, or some other kind of magical, nonexistent thing

Avid consumers of science fiction[54] will recognize that time travel scientists have been attempting to deal with the issue of intense gravity through, well, Science. And what can Science do about a thing that murders people? Why, build a machine preventing that murder!

When it comes to black holes, you're going to need something to counteract all that gravity so that you and your ship aren't instantly spaghettified. We're going to tentatively call that a "gravity drive," meaning some kind of engine that produces antigravity, or nongravity, or antigravitons, or something. You should probably consult a physicist.[55]

We honestly have only wild speculation about what a gravity drive might be, do, look like, or cost. You can bet your ass it's expensive, huge and requires a lot of energy, because black holes

---

54 Most science fiction stories are actually coded communiqués between time travelers. Like that story about the drilling team that boards that asteroid in order to put a nuke inside it? Time Travel Defense Magistrate Bruce Willis was in it—yeah, that was ours. Time Travel Secrecy Advisor Steve Buscemi decided to join the project because he was afraid it was too realistic and had to act crazy to throw people off.

55 Publisher's Note: The authors of this book are not, have never been, and should in no way be confused with physicists.

aren't going to let you use a four-cylinder engine from your mom's old Camry to counter the combined sucking might of the universe, and it may just spaghettify you where you stand for even suggesting such a thing.

They're far away

Black holes aren't located anywhere handy and they don't make house calls. That means a faster-than-light ship and potentially cryotechnology in order to reach one, which basically means equipping yourself for relativity-based time travel into the future in order to turn around and (maybe) use your black hole to get to the past. And then, when you get there, you'll have to avoid becoming pasta as you go through the black hole, if such a maneuver is even possible. There might not even be anything on the other side and there might not even be another side at all. We hope you're feeling lucky.

CONCLUSION

There you have it. You now possess an entry-level understanding of time travel Science. But to turn any of these concepts into any practical means of time travel or accidental death you will require at least one more key ingredient: a time machine.

Chapter 3 will illustrate[56] the five basic categories of time machine and offer you the opportunity to choose the one that's right for you by applying what you've learned here. Or, more likely, choose the one that's right for you by deciding which one looks the coolest.

---

56 No, really: There are illustrations.

3

TIME MACHINES—BUILDING THEM AND INEVITABLY DESTROYING THEM FOR THE GOOD OF HUMANITY

NOW THAT YOU have familiarized yourself with the various means of temporal relocation, we should look in on your pre–time travel checklist:

❑ Procure time travel guide
❑ Discover that time travel is real and awesome
❑ Obtain time machine

That's right—you're just that close. Unfortunately, this third step presents the biggest challenge for the start-up time traveler. It can take several encounters with slippery, fast-talking time machine salesmen and black-market vendors before finding the time machine that's right for you. And that doesn't even take into consideration your affinity for burnt umber, a frustratingly popular paint color.

And although machines were only loosely mentioned in the previous chapter, the authors of this book would like to make a patent endorsement: **When traveling through time, a TIME MACHINE is not optional.**

In addition to the aforementioned benefits (prevention of skin and hair loss, achievement of requisite speeds, looking awesome, wormhole bitch-slap protection), time travel simply cannot be achieved without the aid of a machine.

If you would like to plummet headfirst into a black hole in your underpants, be our guest. If you would like to strap a flux capacitor to your back and see if you can run at eighty-eight miles per hour, go ahead and try. If you, time-traveler-to-be, would like to even begin to contemplate how long and boring relativity travel would be without some kind of near-to-faster-than-light-speed device, promptly hand this guide over to a competent friend or relative who one day might actually derive enjoyment from travel through time.

A machine. No exceptions.

THE MACHINES

Though there are enough time machines built by uncertified mechanics, certified maniacs and unstable experimenters to fill Mogwalth Hu-Man Stadium,[57] nearly all of the models considered just about safe enough for human use fall into one of five categories:

- Car-based
- Booth-based
- Ship-based
- Closed-loop
- Portal-based

If your supposed time machine can't be categorized as such,

---

57 Largest outdoor futbol arena in the galaxy from 2143 to 2201, at a capacity of 2,000,000,013. The thirteen final seats were added late in construction to overtake rival stadium Kim Jong-il VIII Everyone Has A Happy Funtime Stadium on the Saturn moon Titan.

Phil Hornshaw & Nick Hurwitch

caution and thick full-body rubber suits are advised. Familiarize yourself with each type of machine and choose the one that's right for you, or refer back here for operational assistance in the likelihood of an emergency.

Psychosis

It should also be noted, footnoted,[58] and <u>underscored</u> that derangement, psychosis and other forms of so-called "mental time travel" are not, in fact, time travel. They are not even real, depending on your definition of "real" (which, for purposes of time travel, should admittedly remain loose).

The following rumored methods of time travel will not result in actual time travel and may result in admittance to a mental institution or arrest:[59]

- Comas
- Drugs
- Thinking really hard about time travel
- Dreaming about time travel
- Drug-induced dreaming about time travel
- Really hoping you can time travel
- Chemical imbalances of the brain
- DNA abnormalities
- Superpowers
- Being on television[60]

To reiterate: no machine, no time travel. Everything else is imaginary, made-up, the by-product of an unfortunate upbringing, felonious, or some combination thereof.

---

58 Here it is.

59 Machine-based methods of time travel are not guaranteed to prevent admittance to a mental hospital or arrest.

60 Syndication is considered by some to be a mild form of time travel.

CAR-BASED MACHINE
(speeding portal entry)

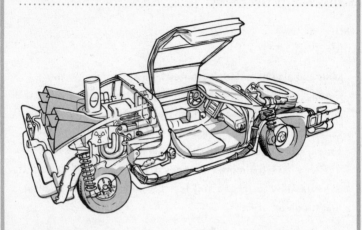

BACKGROUND: One of the most reliable and useful forms of time travel, car-based time machines are the bread and butter of time travelers originating in the late twentieth and early to mid twenty-first centuries. These machines are especially beneficial because they double as vehicles, which means they can be used for standard ground travel, escape, or even hobo-like living accommodations when stranded or down on your luck. A steel-bodied car is preferred to protect against the effects of interdimensional travel. Later car models (late twentieth century on) contain too many plastic parts, which melt easily or break the first time you hit a pothole. Advanced, self-repairing metals developed around 2300 paved the way for a car-based time machine renaissance. Emmett Brown was the first to make a viable car, using his highly stable flux capacitor technology, but he owed much of the precursory comfortable chair and cup holder developments to H. G. Wells's wheel-less, unenclosed early twentieth-century design.

Phil Hornshaw & Nick Hurwitch

INVENTOR: Dr. Emmett Brown, H. G. Wells (easy chair, stationary model)

TIME TRAVEL METHOD EMPLOYED: Wormhole

USEFUL FOR: Eras spanning the Age of Empires to the Robot Apocalypse

NOT USEFUL FOR: Getting around jungles, traversing water, anyplace you're going that lacks roads, times before fossil fuels

YOU'LL NEED:
- Steel-bodied car (preferable)
- Fluxor
- Radioactive fuel source
- Radioactive fuel source containment unit
- Cup holders
- Radioactive fuel source safety gear
- Gasoline
- Doors, murdered out
- Wiper fluid
- Cassette tape player
- Spare tire
- Spare capacitor
- Roller brush for dog hair
- Therapeutic bead seat cover

BOOTH-BASED
(self-contained portal entry)

. . . .

BACKGROUND: Stationary booths are a highly popular model of time machine, though they are far less practical and more difficult to build than their car-based brethren. The technology for building such booths—which eliminate the eighty-eight-miles-per-hour wormhole traversal speed requirement—did not become viable on Earth until the late 2600s. Several alien races developed it considerably earlier but were unwilling to share. Booths have the benefit of blending in during times that actually use pay

phones, and some, like the Time Lord models, can be quite luxurious, equipped with technology that allows them to be bigger on the inside than on the out. However, their lack of mobility in non-time dimensions can make them less versatile and more susceptible to damage and attack from hostile forces than their mobile, car-based counterparts. It is unclear why the Time Lords, an advanced alien race, or people from twenty-seventh-century Earth would choose to disguise their time machines as booths meant to house technology outmoded by 2002, but novelty and inside jokes are widely suspected as being responsible.

INVENTOR: Time Lords, George Carlin[61]

TIME TRAVEL METHOD EMPLOYED: Wormhole

USEFUL FOR: Phone booth–era travel; hiding cool things; secret agent–like stealth

NOT USEFUL FOR: Going anywhere after landing, time travel on a budget

YOU'LL NEED:
- Phone booth / police box
- Phone optics
- Higgs Boson inverse dark matter field regenerator
- Time dilation antenna
- Wormhole traversal shock absorbers
- Wet bar
- Putting greens
- Uranusian metallic synthetic conductive regenerative paint
- Coin return
- Sonic screwdriver kit

---

61 Patent battle ongoing.

SHIP-BASED

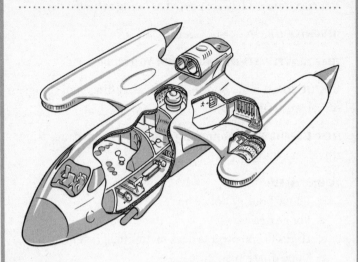

BACKGROUND: Time travel with a spaceship is based on the ability to go as fast as or faster than the speed of light. As relativity explains,[62] travel to the future becomes relatively possible for you the faster you travel—time progresses slower for the people in the ship than it does for people moving at "normal" speeds (like on Earth). This brand of travel is one-way only. It is also possible for ships capable of space travel to navigate black holes or naturally occurring wormholes, but destinations in space and time are virtually impossible to pinpoint, unless someone else has traveled through that same black hole before and lived to tell the tale. This is unlikely. A benefit of the ship-based approach to time travel is that even if you're not using it to time travel, it's still a space ship. Assuming you're adequately stocked with supplies, have a killer holographic

---

62 This should be sinking in by now.

entertainment system and have not yet been boarded by hostile aliens, this is a pretty cool place to be.

INVENTOR: Zefram Cochrane

TIME TRAVEL METHOD EMPLOYED: Relativity, black holes, special-case wormholes

USEFUL FOR: Space travel, travel to the future

NOT USEFUL FOR: Local time travel, travel to the past

YOU'LL NEED:
- Space-O-Dynamic hull
- Faster-than-light drives
- FTL drive power source
- Full-service twenty-four-hour gym
- The bridge
- Escape pods
- Synthetic gravity generator
- Holodeck
- Space torpedo
- Super-helpful, non-murderous artificial intelligence system

CLOSED-LOOP

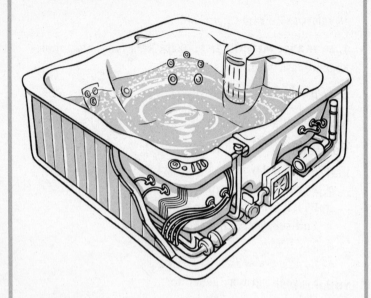

BACKGROUND: The closed-loop machine capitalizes on quantum entanglement to allow for travel back to any point in the past during which the machine (1) existed and (2) was activated. It creates a particle field or allows for particles within the machine to become entangled with their counterparts at the original time location, such that when an object or person enters the machine, that object or person skips backward through time from the present to any point in time since the machine was activated.[63] So, unlike other time machines, a closed-loop machine cannot take you to the future or to any point in time before it was built. This typically means you will not be traveling very far into

---

63 To become even more confused about how this works, watch *Primer*. Frustration and DVD-snapping are common reactions.

the past. And, even if you do, the point in spacetime to which you traveled will remain entangled with the point in space-time that you came from. This makes getting lost in time a lot harder and returning home a lot easier, by which point you will have hopefully learned an important life lesson. The physical appearance of the machine is incidental and the science employed is somewhat confusing.[64]

INVENTOR: Candido Jacuzzi

TIME TRAVEL METHOD EMPLOYED: Quantum entanglement

USEFUL FOR: Cheating the stock market, fixing a single terrorism event, learning to appreciate your life back in the present, relaxing with a combination of heat and bubbles

NOT USEFUL FOR: Traveling back farther than the date of the device's creation, traveling forward, traveling dry

YOU'LL NEED:
- Hot tub
- Bubble/temperature control
- VORTEX™ water feature
- Liquid silicon deadly friction-reducing skin protection gel (in place of water)
- Safe, permanent location
- Failure Clean: the biological bits filtration system
- Cup holders

---

64 Little-known fact: No one ever figures out quantum physics. Ever.

PORTAL-BASED

BACKGROUND: Vehicle-less time travel can be useful under the right circumstances, allowing for the machine to stay in the time in which it was built while its passengers are sent through time for the purposes of adventuring and the discovery of history's secrets. The machine doesn't travel because, rather than facilitating travel through the portal, it merely creates the portal that the traveler can then use (or shove others through). This is usually either a short-term wormhole or a gap in the Quantum Foam through which the traveler is shrunk/deconstructed, then squeezed through, prior to theoretical re-bigisizing/reconstruction on the other side. Portal-based time machines are ideal for sending cyborgs and soldiers to the past on suicide missions, but not so good for returning from the past in order to stop your father's rival from turning your hometown into a den of sin and debauchery. For obvious reasons, portal-based time machines are for one-way travel only.

INVENTOR: Cyberdyne Corporation, ITC Corp.

TIME TRAVEL METHOD EMPLOYED: Quantum Foam, wormholes

USEFUL FOR: One-way trips to the past to assassinate political figures, witness protection programs, time gambling[65]

NOT USEFUL FOR: Getting back home ever

YOU'LL NEED:
- Self-aware robot technology
- Self-regenerative space metal
- Nuclear power source
- Man-invented, machine-improved cold-fusion Tesla coils
- Hologram generator
- Self-aware operator AI
- Safe, permanent location

---

65 See section in Chapter 5 "Time Gambling."

DESTROYING YOUR TIME MACHINE

Now, as you shop around for or attempt to build your own time machine, it is important to keep at bay the human inclination to become attached to material possessions. This is unhealthy and can even be considered a psychological disorder if your time machine arouses certain . . . feelings. More importantly, despite all the adventures you will have together—and all the money you will invest in leather seats and a killer stereo—you will one day have to look your beloved time machine in the unstable power source and destroy it.

Time travel is a dangerous game: as unpredictable as chess and as disorderly as checkers. But the pawns in the game of time travel are real, and the double-jumps take place over unsuspecting timelines. As this guide will oft remind you, danger and the potential to leave your king completely unprotected due to a temporary lapse in judgment lurk at every turn. And, as we will delve into deeply in Chapters 4 and 6, it is possible that time has no refresh button. No new board with which to make up for a poorly played game. The more you time travel, the greater the likelihood that you will one day:

- Screw something up
- Encounter yourself in another time
- Accidentally change time
- Be tempted to purposefully change time[66]
- Murder a witness

And last, but probably first: The more you time travel, the less you'll be inclined to give up your amazing, science-given ability to lead a human revolution against hostile space aliens and shake the hand of Jesus Hector Christ in the same day. And your unwillingness to part with said science-given ability will inevitably

---

66 Which never works out the way you think it will.

lead to at least one of the above, which can in turn lead to a paradox[67] or the dissolution of all existence.

It will be up to you to determine when it is time to destroy your time machine. But the choice is inevitable: either prevent your own temptations, or succumb to them; prevent others from stumbling upon your machine and repeating your mistakes, or enable them; eliminate evidence in the event that actions encouraged by this book ever come up in a court of law, or completely screw us over.

The choice, as we said, is yours.

A Popular Way to Dispose of One's Time Machine

METHODS OF DESTRUCTION

There are several ways to destroy your time machine. More than likely you will have to improvise or use your surroundings to inflict maximum damage in a short time. It is important to remember, however, that you can't leave any evidence behind—and even if you do (which you shouldn't) your machine can't be in any way operational or repairable.

---

67 See Chapter 4, "The Perplexing Pandemic of Potential Paradoxes"

Just in case you have the luxury of choosing the method of your machine's destruction, or you are unsure of appropriate techniques, here are a few ways that QUAN+UM employees have found success when destroying their machines:

- Push it off something up high
  - Airplanes > Cliffs > Buildings
- Cut the wires; pee on the wires[68]
- Blow it up with dynamite
- Fly it into a star
- Strap it to train tracks
  - Be sure to witness horrific train wreck before leaving machine behind[69]
- Bury it beneath a ficus sapling[70]
- Sell it off, part by part
- Crash it into another time machine
  - Pro: 2 for the pain of 1
  - Con: Mushroom cloud
- Drop it into a volcano
- Go crazy with a pipe wrench
- Position it in front of dinosaurs; threaten dinosaurs with fire; get out of the way as dinosaurs trample time machine

Voilà: scrap metal. If you lack the testicular/ovular/asexual fortitude to destroy your machine, or are as yet unconvinced that doing so is necessary, Chapter 4 will enlighten you to the nearly infinite ways that you and your machine will lead to the doom of all existence and, if we are properly doing our job, cause a fear-induced bowel movement.

---

68 May require more pee.

69 Why wouldn't you?

70 You will never again curse them for destroying your plumbing. Also important to note that this one takes time and there is a 0.003 percent chance of inadvertently growing a time travel tree.

4

THE PERPLEXING PANDEMIC OF POTENTIAL PARADOXES

THE PARADOX.

The ruptured Anterior Cruciate Ligament of space-time. The agonizing, cold sweat fever dream of every time traveler ever to crush an arthropod or sneeze on a historical figure of note. The unwieldy, irreparable result of a mistake so seemingly little, yet so indescribably vast, it not only ruins your time-cation but implodes The Fucking Universe as well.

Thus far no time traveler has been so dimwitted as to let a paradox come to pass. If this book serves you no other purpose, please allow it to prevent you from becoming the first.

By definition, a paradox is a statement or proposition that, despite apparently sound reasoning from an acceptable premise, leads to a conclusion that seems senseless, logically unacceptable, or self-contradictory.[71]

---

71 New New Oxford Dictionary, 2043.

Nontemporal paradoxes

This sentence is false. A sentence which if true is false and if false is true. The Homer Simpson Burrito of Doubt. Could God make a burrito so hot that he himself could not eat it? The Chicken or the Egg. Which came first? Answer: Velociraptor.

However, unlike the standard, thought-experiment variety of paradox, the temporal paradox is a good deal scarier. It is fraught with questions, steeped in conundrums, and washed over with an air of no-frills, pants-crapping fear.

Before delving into the various categories of temporal paradox and anecdotes that illuminate the ease with which they are created, we'll start with the basics and allow you a moment to don an adult diaper.

GRANDPAPPYCIDE

The classic example of a temporal paradox is the grandfather paradox, also known as the Grandpappycide Paradox. In it, a moron with the capability of time travel (let's call this traveler "You") uses said capability to go back in time before your grandfather has knocked up your grandmother or that prostitute in Saigon.

While there, you kill Grandpa in cold blood. In addition to issues of murder and family murder and the psychological damage it would cause or require, this scenario creates a paradox. If your grandfather is dead before he makes your dad, then you never could have been born and therefore never could have grown up, found a time machine, traveled back in time, and murdered your own grandfather. Consequently, your grandfather could not be dead, even though you just lobotomized him with a salad fork. One of these factors does not compute.

Phil Hornshaw & Nick Hurwitch

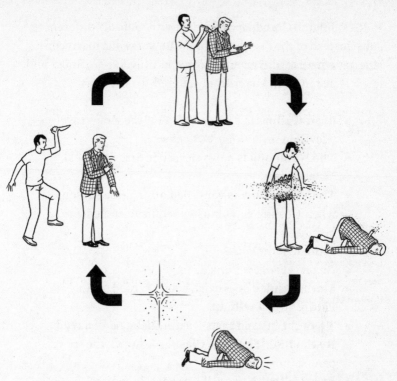

Beyond the obvious questions ("Why the hell would you do that?" or "Couldn't I just kill my dad instead?"), surely you're wondering what would happen if the above scenario came to pass. Unfortunately, there is much more learning that needs to take place on your end before you'll be ready to hear the answer. Suffice it to say that if you are truly hell-bent on avenging your mediocre childhood to the extent that you would murder Pop Pop, please do your best to leave time travel out of it.

THE BUTTERFLY EFFECT

Another classic example of a temporal paradox is known as the Butterfly Effect. The Butterfly Effect occurs when you go back in time and step on the small child version of Ashton Kutcher.

It's similar to Grandpappycide but with radically different results: instead of the dissolution of the universe due to irreconcilable yet equally valid occurrences, the Universe we know and love[72] is instantly and irreparably altered. To wit:

- Ryan Gosling is cast as Kelso on the time travel television series *That '70s Show*.
- **The Notebook** is only released in Spanish-speaking nations.
- **That '70s Show** is cancelled after three seasons when Gosling can't muster requisite comedic relief.
- **Punk'd** never exists.
- No one ever gets Punk'd.
- Maybe **Punk'd** does exist, but it's hosted by Robert Pattinson of **Twilight**.
- **Twilight** instead casts Dustin Diamond of **Saved By the Bell: The New Class** as Edward, the sexy, pale-skinned vampire.
- Even girls with low self-esteem cringe.
- **Twilight** never garners enough success to justify a sequel.
- **No one ever finds out what happens to Bella.** And finally,
- **Demi Moore is available,** assuming you can defeat Bruce Willis in a cage fight to the death.

All right, so all that doesn't sound so bad.

But what if punting pre–*That '70s Show* Ashton Kutcher into a chasm changes more than just Ashton Kutcher–related things? Dubai may never exist, *American Idol* could become a national sport, or you could be Gary Coleman. Who is not only peculiarly tiny, but died in 2010. Paradox.

---

72 Hereafter referred to as "The Universe."

In other circles, the theory of the Butterfly Effect postulates that if you go back in time to prehistory and accidentally step on a butterfly (for example), that seemingly minor change sets off a domino-like chain of alterations throughout the world, which result in a future unrecognizable to the clumsy-footed traveler.

Here the damage done to The Universe is directly proportional to how far back in time you travel.[73] For example, if you go back to yesterday and step in some dog dook, it is unlikely that once you return to the present anything will be different, outside of your need for new sneakers. But say you go way back to a time when the soup du jour is primordial. There you are, on the pristine beaches of Pangaea, just as the first-ever legged-fish demon creature is pulling itself onto dry land. If you kick that slimy toadmonster back from whence it came, you more than likely will have just altered every infinitesimal event to follow and probably even disallowed for the possibility of human existence. At best, we would evolve into merpeople.[74]

It should stand to reason that the further back in time you go, the more careful you should be.

THE FOUR PARADOXES

Thus far you might not have needed that diaper you put on. Most paradoxes, as you are surely seeing, can be avoided through caution, non-murder, and moderate consumption of alcohol.[75]

But don't remove your crap-catcher just yet. Unfortunately, there are many more ways to damage the timeline than Ashton

---

73 This also applies to future travel. The effects of an event or action in the present will have farther-reaching expression the farther into the future you travel subsequently.

74 Worth footnoting: Time travel proves evolution. We think.

75 If you're drunk enough, suddenly things like Grandpappycide and hunting down man-boy Ashton Kutcher seem justifiable.

Kutcher would have you believe. We have carefully devised four different categories of paradox, each trickier and more dangerous than the last:

1. Paradox by Action
2. Paradox by Inaction
3. Paradox by Predestiny
4. Paradox of a General Lack of Information About Paradoxes

The rest of the chapter will describe each of the above variations of paradox in detail, rely heavily on guessing in lieu of actual science, and maybe—if there's time—take a stab at how to avoid destroying spacetime when faced with each.

PARADOX BY ACTION

The primary forms of paradox creation and universe death include the above examples: the Butterfly Effect and Granpappycide. Both involve going back in time and doing something (which better have been accidental[76]) to alter the timeline and potentially kill everyone.

Because we've already covered the basics of Paradoxes by Action, we will now focus on abstinence. And by that we mean both abstaining from changing the past and also from having sex.

As indicated in Chapter 1, the best way to avoid a paradox by action is by not going to the past. But if you are going to the past, the next best way to avoid a paradox is by not going to your own past.[77] If you already are in your own past, or are forced to go to

---

76 For your sake.

77 If you are in a timeline already altered by the effects of time travel, be careful not to cause a paradox by inaction. See below.

Phil Hornshaw & Nick Hurwitch

there by fate, destiny, or a high stakes game of Truth or Dare, do not, under any circumstances, become romantically involved with an ancestor or family member.[78]

If Martin McFly has taught us anything—and he has taught us plenty, and not just about life preserver fashions—it's that your mom may have been, at one time, pretty hot. And if that version of your mom were to meet you, in a gross, drippy, Oedipal, Freudian twist, she might find you to be "baby daddy" material. Scientifically speaking, it may be merely the maternal instinct of your mother unconsciously recognizing her own genes. But as a horny teenager, she could hardly know the difference.

In the film version of McFly's first foray into paradoxical danger,[79] *Back to the Future*, the longer he is in the past, the more he is forced to avoid the sexual advances of his hormone-crazed mother. And the closer he gets to accidentally displacing his father, the more radically he alters the future.

McFly's failure gauge in this instance, which you'll never have, is a family photo from his present that he happens to have with him in the past. As the threads of his family history unravel, McFly's brother, sister, and even Marty himself, begin to fade from the photograph. This is undoubtedly the one fatal flaw presented by the film adaptation of the tale: We are here to tell you that the wrath of spacetime is much more swift and violent than casual erasure from a Polaroid picture, and its fabric is never meant to be shaken. It is of utmost importance (and sickening for the writers to have to point out) that you do not entertain such parental advances, romantic or sexual.

To put it another way: If you in any way prevent the meeting of your parents or the consummation of their relationship in the form of exchanging bodily fluids, it could mean the instant

---

78 What are you, a Nobel Prize–winning theoretical physicist?

79 Which is admittedly totally unreliable.

WRONG   RIGHT

erasure of your existence, which could, in turn, **implode the universe**.[80] Additionally, it's gross.

But fear not. Well, fear a great deal—but also hope some. There is a slight chance, as is the case for young Martin, that some wiggle room exists. McFly's efforts to rectify everything that was changed by his presence in the past doesn't exactly "fix" the timeline—and the events of his life and the lives of those affected don't happen exactly as they once did in the original timeline—but McFly still makes the important parts work. This is known as Marty McFly's Close Enough Axiom. He is even lucky enough (which you won't be) to improve on a few things. His father, for example, now has testicles. And his father's enemy now washes his father's car.[81]

So if you mess up the past—whether it's by killing an ancestor, stomping on a prehistoric bug, or having unprotected inter-

---

80   See section in this chapter "Grandpappycide."

81   *Outtake*: and testicles.

course with your mom—you'll have to fix it yourself. Or at least get reasonably close in your attempt. All it takes is a little elbow grease, some luck, and a hover board. You also may require a mad scientist, Chuck Berry, a sheepdog, and a sleeveless coat/vest. That should about cover anything that comes up.

PARADOX BY INACTION

This second brand of paradox is a bit spookier. It comes into effect when your timeline has already been affected by time travel and your own travel through time is necessary to set up events that have already occurred in your past.

For example: You have a birthday party when you're six. Your parents hire Bobo the Clown to entertain: magic, face paint, balloon monkeys, that sort of shit. On the way to your party, a friend's parent hears on the radio that Bobo the Clown is a murderous pedophile wanted for the armed chainsaw robbery of a local Kids "R" Us.

As the children without chaperones huddle in your mom's dark closet and take turns sobbing into her blouses, your dad tries desperately to catch Bobo on his cell to cancel the appearance.

Somehow, Bobo never shows up. Everyone breathes a sigh of relief, your guests go home with Ninja Turtle kazoos and slices of Seran-wrapped confetti cake, and you hope that these parting gifts are enough penance for their abject terror. So concludes the third-worst birthday you'll ever have.

As it turns out, Bobo the Clown was on his way to your party when he was struck bodily by a decommissioned city bus and ran himself through with his own chainsaw on impact. A fortunate coincidence saved your sixth birthday party and the lives of you and your parents.

. . . Or did it?

You grow up to become a robust reader of how-to guides and

one day find yourself having traveled through time to the exact date of your sixth birthday. While buying a pair of Bugle Boy suspenders to blend in with your surroundings, the Kids "R" Us receiving your patronage is held up by a murderous, chainsaw-wielding clown.

That's when you realize: YOU must murder Bobo the Clown. If you leave it to chance and it wasn't someone else who happened over the clown with a city bus, Bobo will show up at your birthday party, chop up your parents, and feed them to you while your friends watch. By now you can surely connect the paradoxical dots: If you're murdered at six, you can't grow up, can't become a robust reader of how-to guides, can never travel through time, and certainly can never stop that evil fucking clown.

This time you were fortunate. You were wise enough to recognize that Bobo had to die at your hand and, incidentally, that you had always wanted to steal a bus. But that clown nearly had to step on your right big toe for you to come to the realization. Like Denzel Washington in the QUAN+UM time travel training film *Déjà Vu*, who is repeatedly assaulted with clues of his own time travel and fails to make the connection, you won't always be so lucky.

BILL & TED'S MOST EXCELLENT RULE OF TIME TRAVEL

William S. Preston, Esquire, and Theodore "Ted" Logan were a pair of reefer-loving teens from San Dimas,[82] California, whose time travel heroism is often overshadowed by the elder statesman of Suburban California Teenage Time Travel, Martin "Marty" McFly. But their somewhat less well-known phone booth time machine[83] adventures smoothed over some important

---

82 Even in the future, San Dimas High School football rules.

83 For more on phone booth time machines, see Chapter 3, "Time Machines—Building Them and Inevitably Destroying Them for the Good of Humanity."

paradoxical wrinkles and inspired a generation of future phone booth time travelers.[84]

By all accounts, Bill and Ted were staggeringly, mind-numbingly stupid. But they were also totally righteous dudes. If they can round up nearly a dozen historical figures and Winona Rider for a high school history assignment without destroying everything, then screwing around with the past must be possible (a conclusion backed by inductive reasoning).

But what Bill and Ted lacked in brainpower they more than made up for in keen understanding of cause and effect. If I smoke this doob (cause), then I will get superbaked (effect). If I fail this history exam (cause), then I will be shipped off to military school by my asshole father (effect). If I nail this role (cause), then I will be offered the lead in *Point Break*, a future cinematic triumph (effect).

Early in their adventure, Bill and Ted are visited by a slightly-in-the-future version of themselves between trips to the past. The future version of the pair offers Present Bill and Present Ted some vague advice about princesses, interspersed with a chorus of "Dudes" and "Totallys." Other than a complete bastardization of English, this appeared on the surface to be pretty standard time traveling stuff. But when Present Bill and Present Ted began their own time travel to the past, they eventually ended up on the other side of that same exchange, offering "Dudes" and "Totallys" and vague advice about princesses to their slightly-in-the-past selves. And it is there that their brilliance began to shine like a jewel-encrusted Fender Stratocaster: The time-traveling duo had the wherewithal to execute the exchange identically and to give identical advice to their past selves.

Their prior knowledge of the upcoming meeting with their past selves allowed the duo the dangerous opportunity to deviate from the original exchange. If they had deviated, they could have set about on a course that would have led to a radically different encounter with themselves, or prevented the encounter

---

84 Dr. Who is big into reefer.

from happening at all. But, because Bill and Ted experienced it in their past—thus proving it would take place in their other selves' future—any deviation could have meant tearing the ass cheeks of spacetime with an atomic wedgie–like paradox.

This is Part One of Bill & Ted's Most Excellent Rule of Time Travel: If you possess knowledge of actions your future self has taken while time traveling, particularly those that affect your own past, you must execute those same actions with precision.

Sometimes these like executions just happen. Because they already appeared in the timeline, there are typically forces in motion outside your own free will[85] that allow for them to happen again. But you can't let the fact that you know what you're supposed to do influence a deviation from your course.

Now, Bill and Ted, in their infinite righteousness, took this one step further. They realized that if they had to do what their future selves did, then they could also do something in the present to influence (or even help) themselves in the future.

While busting out their historic superbestfriends (Socrates, Billy the Kid, and Abraham Lincoln, to name a few) from a San Dimas city holding cell, the time-traveling pair realized they needed the key to the cell. Munchies would also be clutch.

They agreed—in the present—to go back in time—from the future—to steal Ted's dad's prison cell keys[86] and hide them behind the sign in front of the police station.

Suddenly, instantly, miraculously, the keys were there—right behind the sign as they'd agreed. Bill and Ted merely needed to remember, they reminded each other, to travel back in time later, after their bodacious history presentation, to steal Ted's dad's keys and place them behind that sign. Otherwise the keys wouldn't be there. But of course, they were there. If they had forgotten or were somehow prevented from traveling to their

---

85 Known in physics as a "weak force."

86 Which of course he takes home with him every night.

past from the future, the keys could never have been there in the first place, and we would have yet another paradox on our hands.

This is Part Two of Bill & Ted's Most Excellent Rule of Time Travel: If you have the reasonable means to time travel and agree in your present to travel to the past from your future, the effects of such a journey will express themselves immediately.

Of course, if you make such an agreement and nothing happens, things don't look so good for you and your future time-traveling endeavors (i.e., in the immediate future, something horrible, such as death, seems likely). To keep safe, you could try hiding in a dark place, or you could opt for offing yourself now while the implement of your death is still a matter of choice.

PARADOX BY PREDESTINY

Up until now, we have primarily dealt with repercussions of travels to the past. But what of the future? Is it possible to screw up your life and the lives of others to an almost incalculable degree by traveling forward in time?

As United States Secretary of the Treasury Alexander Hamilton will tell you: yes. Or at least, that's what he would tell you had he not been shot in the abdomen by Aaron Burr, his political arch-nemesis. While many know Hamilton as one of our formerly great[87] nation's founding fathers, he was also one of time travel's founding second cousins.

And though Hamilton lived out a rare life in which he was a pioneer three times in three completely unrelated fields,[88] his untimely demise by duel was never meant to be. It was Hamilton himself who caused his death by gunshot wound, the result of a man who had learned of his own fate before he had seen it through.

---

87 Not so great following the "No Pants President" election gimmick of 2088.

88 Little known fact: He was technically the first champion bowler.

ALEXANDER "BEAR ARMS" HAMILTON AND THE PREDESTINY PERTURBATION

In 2023, an early time-travel test and historical reconnaissance mission was conducted. The mission sent its testee, the lab's intern, to a period sometime shortly after the birth of the United States of America. The time machine returned to its lab with the intern, a college sophomore named Ricky, flopping out its open door and falling wet, bruised, and shivering to the floor. Stepping out behind him a moment later, shirtless and belligerently drunk, was Alexander Hamilton, one of the nation's most notable founders.

Though Ricky wouldn't explain the circumstances, or even meet Hamilton's eyes, it appeared that through some drunken fisticuffs, Hamilton had accidentally boarded the time machine after stepping out of a pub into an alley to urinate during a night of heavy Philadelphia drinking. There he'd found the young bespectacled intern, Hamilton said, spying his manly parts and had given poor Ricky what for. Ricky, accidentally soaked in blood and urine, had returned the time machine to its rightful era in a terrified panic, bringing Hamilton along with it.

Interestingly, while the scientists recognized Hamilton, none but Ricky was aware of his true significance as a historical figure. While Hamilton had gone on to do many more important things in the history of the United States after that night of drinking in the original, Ricky-free timeline, because he stepped into the time machine, he effectively vanished from history from that point forward. The Universe, it seemed, had instantly corrected for this as well: the lab's operators and everyone else back in Ricky's time only knew of Hamilton's exploits up until the point of his disappearance. Only Ricky—the sole remaining member of the original timeline—knew the truth about his passenger.

With history in a confusing state and further testing clearly required, merely shipping Hamilton back to his proper time was not an option.

Phil Hornshaw & Nick Hurwitch

Much to the surprise of the unnamed experimenters,[89] Hamilton was, after a hot shower and some hair of the dog, rather taken with the future. Being a man of letters and from a time without Costco, air-conditioning, or flip-flop sandals, he was fascinated by every minute detail of the world around him.

Hamilton thusly became one of the first time travel enthusiasts and a willing participant in several dangerous experiments. His enthusiasm was as unexpected as it was rare. In a time when temporal guinea-piggery was referred to by experts and historians as "foolhardy," "a death wish," and "good for the future of humanity only if you hate humanity," Hamilton marched boldly forward. He was, among other things, principal in data collection, the summarization of various epochs and the incorporation of working toilets into time machines.[90]

But his passion remained politics. He spent his free time watching Fox News, C-SPAN, and CNN, often shouting at the television as though a participant in a one-man town hall debate. Hamilton was apoplectic to find not only that his political party no longer existed, but also that modern politics had completely misinterpreted the Second Amendment.

In reconciling nearly three hundred years of American political history, Hamilton eventually befriended Ricky the Intern and, through him, learned that Hamilton had lived a long, full life in the original timeline. Hamilton was instrumental in the formation of our democratic

89 It probably wasn't us.

90 See Chapter 3, "Time Machines—Building Them and Inevitably Destroying Them for the Good of Humanity."

republic and had helped to maintain gentlemanly order in the eighteenth century.[91] He also learned that he lived to the ripe age of 73 and died in a Boston seafood restaurant, where he choked on a rather unfavorably sized hunk of shrimp.

The fact that he had Ricky's historical knowledge at all told Hamilton what he had known deep inside all along: He would one day have to return to his own time and fulfill his duty to his country. And so, once the scientists and Ricky, who had now become his close friends and bowling teammates, had collected sufficient time data and won the Oakland Valley Bowling League title,[92] Hamilton climbed onto his time toilet and bid the future adieu.

The life he finished out was what you will now find in the history books: that of a lifelong politician and dutiful family man;[93] largely the same man he had been before he left. But other than a newly realized aversion to shellfish, there was one enormous change in Hamilton: He had become fearless, adventurous and damn near dangerous. Of all the fatuous, impossible and exciting things that he had seen and experienced in his time as a traveler, none had affected him as much as learning the details of his own demise. Hamilton believed he was invincible, because he knew when and how he'd die.

In the time between his return to the eighteenth century and his death, Hamilton ate undercooked meat, rode bareback,[94] volunteered to fly Benjamin Franklin's kites, was seen in public without his wig, befriended vicious native tribes, petted strange dogs he didn't know, and challenged Aaron Burr to a duel. And

---

91 He was known to have slapped more than a few fellows with a glove.

92 The team fell into tumultuous ruin soon after his departure—not just anyone can routinely bowl in the 280s.

93 With the glaring exception of his rather public affair, subsequent blackmailing and the resulting damage to his reputation. That, too, was unforeseen by Hamilton.

94 On horses.

not just any duel, but a duel atop the same ridge where Hamilton's eldest son had been killed in a quarrel of gentlemen two years prior.

Hamilton had convinced himself that the duel was a certain victory. After all, he knew how he died—years later, in a Boston restaurant, choking on a damned hunk of boiled shrimp. There was no sense in not defending his political honor, especially if he could kill an enemy in the process.

Unfortunately, the only gun Hamilton had used in some years—the shotgun[95]—did not fall within bounds at gentlemanly contests. He was forced to use a shoddy, era-appropriate flintlock pistol. He lined up, fired and missed—horribly. The shot struck a tree branch high above Burr's head and allowed the opponent to take careful aim in return. Hamilton was shot just above the left hip. The bullet shattered two ribs and ricocheted through his organs. Hamilton slowly bled out, cursing fate with his dying breath.

Hamilton's knowledge of his own demise had affected his judgment and changed history. This is the lesson time travelers must learn from Alexander Hamilton, time travel pioneer: Knowing too much about your fate can cause your fate to stop being your fate, with an entirely new fate fating you as fate runs its course.

CAUSATION

Unlike traveling to the past, you will never return from a trip to the future to find your time of origin altered. The universe will still be intact, and every detail of your mundane life along with it. But one thing will have changed unequivocally: you. And not in a metaphysical way, like a summer in the south of France. And not in a sexual way, like getting down with an older woman (or

---

95 See section "SURVIVAL GUIDE: UNIVERSAL RULES AND ADVICE: WHAT YOU SHOULD BRING, 'Boomstick.'"

man [or hyper-intelligent ape]). The changes you will have undergone, the ones that truly matter, are mental.

The future holds myriad exciting and unpredictable possibilities. Will humanity ever achieve peace or solve world hunger? Find cures for AIDS and cancer? Develop new diseases in their place? How many PlayStations will be made before Japan is overrun by its own pet robots? Are the Ninja Turtles angry that we genetically engineered them for our own entertainment, or do they live comfortably in the sewers of New York, defending America from evil ninjas? Does Dick Cheney's heart ever fucking give up? All great questions, many of which we address in the survival section of this guide. And though we know a great deal about what lies ahead for our meager species, in its finest details the future is a smidge difficult to pin down.[96]

The most interesting possibilities for the future, those that your narcissism has likely been salivating over for the last three pages, are personal in nature. The questions you're really pondering are: Will I be rich? How many kids will I have? When I die, will my corpse be jettisoned into space or repurposed as garden fertilizer? How rich will I be, exactly? Will I convert to a polygamous religion in order to wed at the same time the several supermodels attracted by my vast riches?

The answers are all there, intrepid traveler. And therein lies the potential for predestiny paradoxes.

Seeing what is going to happen based on your current course of action in the present, regardless of your mental mettle, is going to have an effect on what you do in the future—assuming you ever make it back to the present at all.[97] If you like how rich you are in the future, chances are you're going to do your best to maintain a course to riches. But even this decision, one that assumes no change on your part, could change the decisions you would have

---

96 See *The Empire Strikes Back*, Yoda: "Difficult to see. Always changing, the future is."

97 See survival guide section "Robots," survival guide section "Space Travel."

made without knowing that you would one day be rich—and those changed decisions could, in turn, prevent your own wealth. If you find yourself dead in the future, tragically unaccounted for in the aftermath of a Manchester United football defeat, left to rot in a ditch that reeks notably of cabbage, chances are you're going to start doing whatever it takes to avoid such a fate once back in your original time. But—once again—the changes you enact based on your knowledge of the future (as was the case with Mr. Hamilton) could be precisely what leads to your grisly demise.

So you see, the paradox here is one less injected with potential universe implosion and more lathered in your own weak mental state and human second-guessing. While the past has already happened, and is therefore easier for us to chart potential alterations within, the current perception is that the future is in constant flux. Your knowledge of what will happen becomes part of what will happen. Perhaps seeing the future is exactly the reason why the future is as the future is. In fact: Bank on it. Or don't. One is probably correct.

What to do? The first solution is rather simple. As with travels to the past, where interaction with your Past Self and Past Self-related events should be avoided at all costs, do not visit your own future. The potential for damage is too great, and we wouldn't want you to preemptively regret a decision like marriage or hair plugs.

But if this tale of caution is too much for you, or you just don't see the point of future time travel unless you can take a peak at your own destiny, try and forget. Knock yourself out with a brick, fall down some stairs, or allow a licensed hypnosis professional[98] to hypnotize you into believing it was all a dream. But don't think for a moment that you can see what lies ahead for yourself and go on living as though the images aren't permanently seared against your mind's eye.

---

98 QUAN+UM does not endorse Hypnotoad or Hypnotoad-related products.

You are your own worst time traveling enemy.[99]

PARADOX OF A GENERAL LACK OF UNDERSTANDING OF PARADOXES

As we've read, the threat of totally mucking up time is never really gone, as any action taken in the past (or future) could result in a paradox. That, of course, totally sucks.

But as far as anyone knows, paradoxes remain a merely theoretical consequence of squashing the wrong bug, kicking the wrong dog, or fixing the wrong World Series. After all, humans have been walking the temporal tightrope and dancing in the multiversal minefield for X^{100} years and we haven't all imploded yet, right? We're all still enjoying ice cream, hydrogen bonded with oxygen, zombie movies and an inflated sense of superiority over domesticated animals and a few tigers.

The authors of this guide hate to admit it, but our information on paradoxes is extremely limited. Countless travelers could have encountered, repaired, or fallen victim to paradoxes—but we just don't know about it. And this is largely the result of a great many time travelers epically failing to properly document their epic failures.

Consider by way of example the legend of time travel legend John Connor. Beginning in the early 1980s, Connor and his kin were the subjects of numerous run-ins with time travelers. But the information we have about these encounters and their resulting effects on time and The Universe remains sketchy at best.

---

99 Other than Hostile Time Travelers. See Chapter 6: The Worst Casetime Spacetime Scenario.

100 Take the year you were born, multiply it by 15, divide that by -1, subtract the total from 2177, and then come to the conclusion that a measure of time when referring to time travel makes almost no sense, on account of you can travel to whatever time you want and time travel for an indefinite amount of time. Let's just say, it's been a while.

The Connors were routinely afflicted by time travelers in pairs: One traveler was, without fail, some sort of cyborg sent back in time to eliminate Connor because of the role he would play as resistance leader in some human-robot war in Earth's future. The second traveler was always some brand of protector, sent by Future Connor to protect his past self.

The first time a pair of these travelers traveled to the past, they dealt with Connor's mother, Sarah. The cyborg failed to kill Sarah and prevent John Connor's birth. The protector, a human, actually ended up banging Sarah during the course of "protecting" her and, by a crazy twist twistier than several twists twisted together, became John Connor's father.

Unfortunately, Hollywood fiction eventually co-opted the Connor story and further twisted it beyond all recognition. First off, when Kyle Reese, the protector from the future, ended up being John Connor's dad, that should have been a major paradox event. Think about it: John Connor sent a guy back in time who ended up fathering John Connor. The future was the antecedent for the past.

That doesn't jibe with the whole "causality" thing we have going in the human conception of physics. Effect follows cause, future follows past, and the future has to be caused by the past in order to occur. It can't happen the other way around.

But it gets weirder: In 1992, another pair of time travelers showed up to screw with John Connor's young life. Again, the evil traveler failed to terminate Mr. Connor. And while there was little to no known boning during this second trip, the protector from the future this time around did help Connor and Sarah destroy the corporation and the technology that was to lead to the robot uprising in the first place.

So a traveler from the future, who was also a robot, was instrumental in the destruction of the robots in the future. Because they were all destroyed, there shouldn't have been any future robots and there should have been no robot protector. Which means, by extension, there should have been no robot war, and therefore there should have been no reason for Future John Connor to send a guy back in time who would bang his mom in the past and beget John in the first place. With no robots, John shouldn't have existed, which would have meant that there would have been no one to destroy the robot company and prevent the robot war.[101]

In essence: The entire universe should be a sucking black hole thanks to crazy robots.[102]

But we're not. And what's worse, the Time Travel Community is not at all sure where the information gets faulty. We all thought it would make a great movie, but in a rush to write the script,[103] everyone forgot to write down the original story, and now all anyone can remember is that Arnold Schwarzenegger was such a badass in the movie, it only made sense to allow him to actually destroy California by electing him robot-governor.

Does free will exist, or is every choice predestined? Will a paradox erase all of existence or is spacetime self-correcting? What about multiple offshoot universes created by altering timelines?

It seems the list of volunteers willing to go discover the answer and see if The Universe blinks out of existence is pretty short.[104]

---

101   We had one of the interns draw this up on a floor-to-ceiling whiteboard, stare at it and try to figure it out. He had a seizure. It makes that little sense.

102   Robots haven't been good for much of anything since the development of 2001's Roomba, which could vacuum an entire house and avoid running into things. Now *that* was a classy robot.

103   There was a whole series of time travelers jumping back to minutes before the sale of that script in order to one-up one another. There were several malfunctions, and a few deaths. A guy named James Cameron ended up coming up on top after the Time Travel Congress's "Great Inter-Temporal Terminator Script Contest" Act of 2244.

104   The List of Volunteers Willing to See If They Can Make the Universe Implode: 1. Ricky the QUAN+UM Intern (college credit contingent).

SCIENTIFIC POSTULATES OF WHY WE ALL STILL EXIST

Discerning readers probably have a lot of questions, even after that whole section we just gave you admitting we have no freaking idea why we all still exist. Well, Time Travel Science has your answers: in the form speculation, rumor, hearsay, conjecture, flat-out lies and theories with no backing in actual Time Travel Science.

With so many imbeciles bouncing around in time potentially screwing things up, you would think we'd be experiencing more problems in general. After all, any jackass with $15.00 can eventually figure out time travel (provided she or he exists sometime after 2013, has the proper theoretical education, and possesses no real appreciation for the sanctity of life and is therefore able to experiment freely).

The following theories attempt to explain how this is all possible and why we haven't all been paradoxed into oblivion.

THE UNIVERSAL THEOREM OF THE UNIVERSE PREVENTING YOU FROM DESTROYING SAID UNIVERSE

One unpopular theory[105] is pretty simple: Time takes care of itself. You can try to create a paradox all you want, but Time knows what it's doing, and Time is smarter than you, since it has been around since the beginning of Itself, which is also before you were born, Smarty Pants.

Recall the Grandpappycide Paradox example. The steps of that paradox are pretty simple—go back in time, kill grandfather.

But the actual performance of said paradox is much more complicated than the description of the scenario lets on.

---

105 Unpopular because it removes that whole epic doomsday thing from the equation, which really takes a lot of the danger, excitement and ambient forbidden eroticism out of time travel.

For example, think about the logistics of tracking down and murdering one's own grandfather. It's not exactly easy. For one, he will look completely different from the man the time traveler may have known during the traveler's life. And just because you know a time that a person lived in does not mean you'll know any specific place that you might be able to find him.

But for argument's sake, let's say a traveler does manage to track down his or her own grandfather, positively identify him, and find a chance to attempt his murder. Any number of things could go wrong at this point. What's more, the Universal Theorem states that they will go wrong.

Say the time traveler takes aim with a gun at the grandfather. The theory states that something seemingly random will intervene. The traveler could find him- or herself emotionally unable to pull the trigger. The gun could misfire or jam. The grandfather could escape, or survive the wound. Richard Alpert could show up and do that magic hand thing.

And in an extremely unpleasant twist,[106] The Universe might wind up and bitch-slap the paradox-inducing traveler by dropping a piano on his or her head, making the gun explode in the traveler's hand, or giving the traveler a well-placed embolism. The possibilities are endless, but the theory states that if you think you can screw with it, Time is not on your side.

THE ELASTIC, FORMFITTING WAISTBAND OF TIME

Similar to the Universal Theorem of The Universe Preventing You from Destroying Said Universe, this theory provides that spacetime reacts to any changes in such an instantaneous and complete manner that no one notices any change at all.

Just as someone well versed in obesity wears stretchy pants

---

106 Although not unpleasant for us—screw you if you're trying to erase all existence. All existence includes us and our (your) money.

Phil Hornshaw & Nick Hurwitch

to a Thanksgiving feast, so, too, is The Universe ready to adjust to alterations in the timeline under this theory. When a time traveler changes things, the changes happen instantaneously everywhere, in every time, with no requirement for antecedents.

Once again, take Grandpappycide for example. Suppose a time traveler did murder his grandfather—the grandfather would die and the time traveler would cease to exist. Instantaneously, the ripple effect would streak through the timeline, eliminating the traveler and, more than likely, his whole family. His very existence and all memories of it would be vanquished from every atom in the 'Verse. Time's "elastic waistband" would expand or contort in such a way as to absorb any and all paradoxes.

The belief here is that there is plenty of wiggle room to paradox anything and everything with zero consequences. That means you can go around murdering hundreds of grandfathers—to Time, it ain't no thang. The Universe was here before you went on a geriatric killing spree, and it'll keep on keepin' on with no regard to you and your cause-and-effect beguiling.

To a degree, this makes sense: It takes more than one idiot to destroy a universe, after all. On the other hand, this is an awfully liberal interpretation of the General Human Conception of Everything. It basically states that the Law of Cause and Effect can't be broken because la la la la la can't hear you not listening la la la la la.

Seeing as most physicists believe The Universe's maturity level to be above that of a four-year-old, the Elastic Waistband Theory usually is disregarded.

TIME, PART 2: THE REVENGE

All great things deserve sequels. In the case of time travel, the Timeline Sequel Theorem[107] imagines time as a single line moving

---

107 Also known as the "Die Harder Postulate."

in one direction, such as left to right. The interference of a time traveler in the past—any interference, including just being in a time the person shouldn't be—creates a new or "divergent" timeline.

Think of it like this: Imagine you jump back in time from 2000 to 1990. In 1990, you advise your mom to buy stock in Google during the next decade, since it'll be all the rage for fifty years, before the robot overlords use the massive store of information and Google Earth satellites to pinpoint and enslave all humans.

Mom buys Google, Mom gets rich. In the original timeline, say "Timeline A," Mom was poor. In the new timeline ("Timeline B"), Mom gets rich because of the interaction of you, Time Traveler X.

Under the Sequel Theorem, everything on Timeline A after the interaction of Time Traveler X with Mom in 1990 would be different than what you remember, even if it's only by a little bit[108]—thusly becoming Timeline B. In order to avoid paradoxes, Timeline B is a sequel timeline to A—A caused B, but A after 1990 is not a part of B.

So any time a time traveler visits the past, it's like a new *Rocky* movie that takes the place of the original, superior *Rocky*. And once you've seen Rocky fight Mr. T, you can't un-see it: You can never return to Timeline A. Timeline A is gone—you killed it. Now you're stuck with Timeline B. And eventually—maybe—the dreaded Timeline Q.[109]

THE MULTIVERSE THEORY OF BALKI TO YOUR UNIVERSE'S COUSIN LARRY

The television show *Perfect Strangers* doesn't have a lot to do with this theorem except that it makes a good example, and we

---

108 For example, if you stepped on an ant, Timeline B would only be different because it would have one fewer ant, but it would still be fundamentally *not* Timeline A. Timeline A has one more ant than Timeline B.

109 For "quixotic."

never pass up an opportunity to talk about 1980s sitcoms.[110]

In the TV show, Regular American Larry's distant, Eastern European, sheep-herding cousin, Balki, comes to live with him in New York. Balki is a little like Larry in some ways, a lot unlike Larry in many others. Yet they are from the same family.

The Multiverse Theory is a lot like the relationship of Larry and Balki. It suggests that every time a decision is made by a thinking individual on the Planet Earth (or any other planet, for that matter), there is a New Universe created. All those other universes are similar and made up of the same stuff but are ultimately different—Balkiverses.

This New Universe, or "spin-off," is almost exactly like Our Universe—except that in the spinoff universe, that one decision went the other way.

Say you flip a coin and it lands heads. Multiverse Theory says that at the moment the coin landed, in another universe it landed tails. And this applies to everything. In fact, the theory allows for an infinite number of alternate universes: one for every infinitesimal change, possibility and decision, multiplied by every other infinitesimal change, possibility and decision.[111]

It's a convenient theory when applied to time travel. Again, we'll take the example of Grandpappycide. When our time traveler travels backward in time and murders her own grandfather, nothing happens. There's no spontaneous implosion, no stretching of Time Pants, no dissolution of existence. The time traveler stands there, totally befuddled, wondering to where her or his paradox disappeared.

But the reason there's no paradox is because no laws of causality have been broken. While the grandfather lies in a puddle of his own bloodsauce and brainchunks on the floor in front of our disturbed traveler, that grandfather is not actually the time traveler's grandfather. The dead grandpappy is a copy of the time

---

110 As all humans are painfully aware, culture and entertainment were at their all-time historical peak in 1986. It's thousands of downhill years after that.

111 See section in Chapter 2 "Quantum Foam."

traveler's grandpappy, and the murder occurred in another universe. This is the one of many infinite universes in which this man never becomes a grandpappy because he is murdered by someone in future clothes.

Under the Multiverse Theory, every time you travel through time, you're actually just traveling to another universe. Because there are an infinite number of other universes, they can exist at all times. But many are crazy, weirdo, horrible versions of Earth, because Infinity is a big number and it thinks discrimination is wrong.

Conveniently, Multiverse Theory makes it impossible to implode this universe or any other by way of accidentally creating a time paradox. Inconveniently, it actually renders all time travel useless, because the alternate times a traveler is visiting are actually other universes. And these universes could be similar or different to the time traveler's universe in any number of huge or tiny ways—but in essence, it's not the traveler's past or the traveler's future. So it's totally unreliable as a source of information.

Side note: It may also be nearly impossible to return to the universe of origin once traveling has taken place. Infinity takes up a lot of room and it's easy to get lost, especially because there are no good maps of the Multiverse. Be aware that the Giant Man-eating Poodle universe might be as close to home as you can get.[112]

LET'S REVIEW

Long, long, long story short—you should really try to avoid paradoxes.

You can avoid paradoxes by doing the following:

- When traveling to the past, never kill family members.

---

112 Recall Marty McFly's Close Enough Axiom from earlier in this chapter.

Phil Hornshaw & Nick Hurwitch

- When traveling to the past, never change anything.
- Never travel to the past.
- When traveling to the future, try not to see what happens to you.
- If you see what happens to you in the future, and you don't like it, try not to make it happen by trying to make it not happen.
- Never travel to the future.
- If you decide in the present to use time travel in the future to do something in the past, make sure you do it, or you'll create a paradox by not doing it.

However, you may not need to totally fear paradoxes, because:

- The Universe might fix it for you.
- The Universe might just pretend you never existed so it doesn't have to deal with you.
- You might only create a less-good sequel timeline that everyone wishes hadn't been made.
- You might only be screwing up someone else's universe.

But luckily, at this juncture, we don't quite know. If we did, we are willing to hypothesize that you wouldn't be reading this guide and that time travel would be outlawed in several states and municipalities at minimum. And though we here at QUAN+UM are all for science-i-ness and learning, in the instance of paradoxes the ends do not justify the experimental means. So please . . . exercise temporal dislocation cautiously and responsibly.

5

THE TIME TRIALS OF TIMEBATTLE

THIS CHAPTER IS likely the one you have been waiting for (or skipped to early, ill-advised though it may be) and focuses on the single most important and practical aspect of time travel: hand-to-hand combat.

If you were beginning to think that traveling through time would be akin to a day trip at Busch Gardens—and that access to the ruinous practices of Egyptian slavery, American slavery, Robot and Dino slavery would be easily granted from the safety of a tour-guided tram car behind four inches of bullet-proof Plexiglas—well, you probably can't afford those tours. Not to mention, you have to wear those embarrassing headsets and are typically not even allowed to touch the slaves.

Though unpredictable, your life up until now has been, for all intents and purposes, a straight line. But as soon as you step foot into a time machine, or accidentally stumble into a rift in the spacetime continuum, your life will become a great deal more complex, like a tuft of yarn played with for hours by a self-entertained house cat. Then eaten by that self-entertained house cat.

Then puked back up by that self-entertained house cat as a hair-ball—still made primarily of yarn but now also containing chunky, unidentifiable bits and the remains of several mouse and bird carcasses.

As soon as you time travel for the first time, all conceptions of beginning, middle and end, past, present and future go out the four-inch bulletproof Plexiglas window. Who knows what times you'll visit, on how many occasions, or how your travel through time will affect your own timeline despite the best efforts of QUAN+UM and its affiliates?

It is only a matter of time before one or both of the following occur:

1. You encounter another time traveler.
2. That other time traveler is a past or future version of yourself.

The assumption that your time travel would take place in a one-man playground where the one man is you and the playground is the Universe is but a foolish daydream. If someone of your physical and mental pedigree can come to possess this guide and also gain access to one or several means of time travel, then the only safe assumption is that literally anyone else, at any time, in any place, must have an equal or greater chance of accomplishing the same.

Imagine, if you will, the yarn hairball that will become your life: tangled, messy, enigmatic and odorous. Now imagine that same yarn hairball entangled with yet another yarn hairball, from another or perhaps the same cat. And another yarn hairball. Yet another. You are now surely seeing (and smelling) what a confused, unmanageable mess your encounters with other time travelers will create for yourself and, much more importantly, for the rest of us.

To prevent your stinky hairball life from combining with their own, more often than not, other time travelers will want to fight you. And if that other time traveler is another version of you, the threat is exponential—the ensuing battle more vicious, more dangerous, more inevitable.

Chapter 5 will teach you, unequivocally, how to be self-defeating.[113]

HOW TO TELL IF YOU'VE ENCOUNTERED ANOTHER TIME TRAVELER

Before we get to effective attack combinations and whether or not you should wear protective armor, we must first teach you to identify hostile time travelers; which means, before that, teaching you how to identify time travelers.

If you suspect someone you have encountered or seen might be a time traveler:

1. Follow the person discreetly. See if he or she ducks into any poorly lit alleyways or looks shifty-eyed over his or her shoulder before entering a dimly lit storefront.
2. Hire a private detective to report back.[114]

---

113 The last thing you want is to be engaged in an eternal sissy fight because you never learned how to kick an ass.

114 So long as you're sure the private eye is not a time traveler.

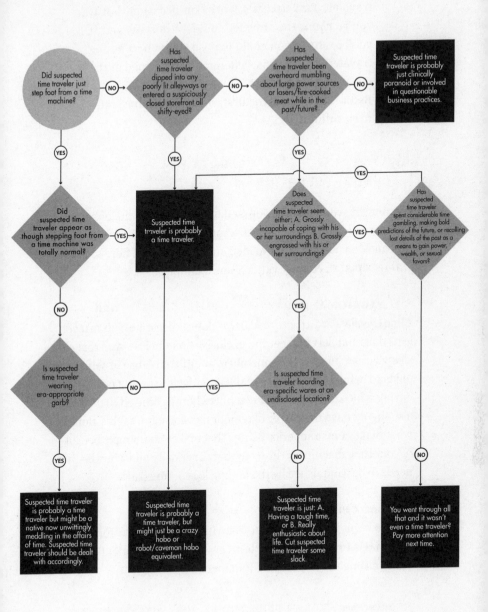

3. Ask the suspected time traveler if he or she is a time traveler. This actually works on occasion, but if you're right, the traveler could be hostile, and if you're wrong, you could expose yourself as a time traveler, or otherwise a heretic to be burned at the stake[115] in certain eras.

4. Before attempting the first three suggestions, refer to this flowchart:

HOW TO TELL IF THE TIME TRAVELER YOU'VE ENCOUNTERED IS HOSTILE

You may be wondering to yourself, why, just by virtue of being a time traveler, said time traveler should be approached with caution and considered robot-armed and dangerous. There are several reasons, all equally valid. Some examples:

MEGALOMANIA

Time travelers are a pretty exclusive club. A disparate club that collects dues and never meets, but exclusive nonetheless. And that exclusivity can inject into its members an inflated sense of self. This tends to manifest either as a really great feeling or as a power trip during which a club member convinces her- or himself that no one else should have the power of temporal dislocation, as this alone is what makes him a superior being. That makes you, simply because of your time machine, a threat to that superiority and a hurdle that needs to be vanquished by the flailing legs of timebattle.

BAD CHILDHOOD

Some people just need to see a shrink. For better or for worse, QUAN+UM has yet to be able to enforce psychological test results in time travel passport issuance.[116]

---

115 See section "SURVIVAL GUIDE: MIDDLE AGES: WITCHES AND WIZARDS."

116 Time travel passports also not yet enforced by QUAN+UM.

Phil Hornshaw & Nick Hurwitch

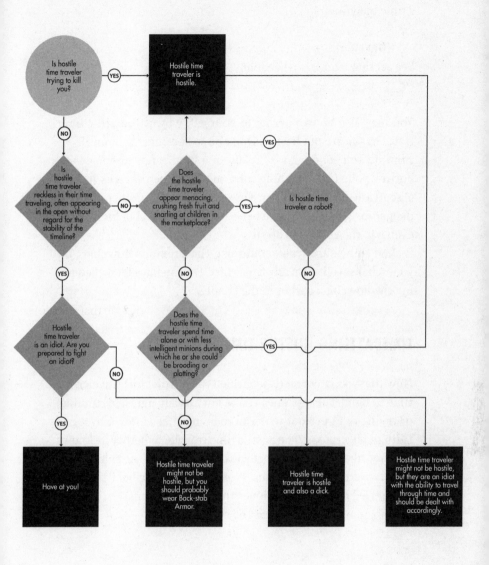

PERSONAL VENDETTA

Be careful whom you piss off; especially if whom you piss off has a time machine.

GREED

See section later in this chapter "Time Gambling."

You may also be wondering to yourself, why, if you are a time traveler, you are not hostile. Don't be so sure. Just wait until you run into someone who is "really in a bind" and "just needs to borrow your time machine for a minute," then returns it with a scuff on the hatch door. Or until you see another traveler dangerously exploiting the benefits of time travel before you've even had the chance to do so.

Just to be sure, you should use the previous flowchart on yourself, as well as on any time travelers you have identified using the flowchart earlier in the chapter:

TIMEBATTLING HOSTILE TIME TRAVELERS

Now that you've correctly identified the hostile time traveler, it's time to fight. This assumes you're not already fighting, running, or trying to figure out why you're being chased down by a guy riding a triceratops. The hostile time traveler, whether he knows it or not, already knows what you don't: There are no rules on the path to victory with no rules. To even the playing field, there are some tried and true techniques that you should master, practice in your spare time, or at least be aware of:[117]

---

117 As they may be used against you.

- **SLAPPING**—Your first and last line of defense against hostile time travelers and children under the age of twelve.[118]
- **AVOIDANCE**—You can't be fought if you can't be found. If you are the lesser timebattler, or are simply a hippie with weak arms, you can often come to a default victory by seeing to it that the fight never takes place. Some tips:
  - Don't go anywhere you've been before or that you've indicated to your enemy you might like to visit someday.
  - Don't go anywhere your enemy has been before or has indicated to you he or she might like to visit someday.
  - Don't go anywhere completely unlike a place you or your enemy have been or might have indicated a desire to visit someday, as most time travelers are well versed in techniques of reverse psychology and dumb luck.
- **SHOTGUN**—If you brought one (which you should have[119]), the shotgun is a much better deterrent than slapping and much more effective on the thin, supple skin of children under the age of twelve. However, this is an Item of Last Resort, and you should be aware that brandishing your shotgun can be as effective (and sometimes more so) as actually using it. Some tips:
  - Don't fire unless you have to.[120]

---

118 Unless said twelve-or-under-year-olds have already hit puberty. Then they might be bigger or stronger than you, at which point slapping may prove ineffective.

119 See section "Survival Guide." All of it.

120 You have to.

- Don't use in an era without shotguns.[121]
- If used in an era without shotguns, be sure to silence witnesses.[122]
- If you don't want an enemy to follow, aim for the legs.
- If you don't want an enemy to follow ever, aim for the face.
- If you don't want an enemy to follow you for probably ever, aim for his or her time machine. Which brings us to . . .

- **SABOTAGE**—This is the trickiest, but perhaps most effective, method of claiming victory in a timebattle. Your nuclear-powered, lead-plated life preserver of time travel is, of course, your time machine. The same will be so for other time travelers who wish to take your lunch money and life. Some methods of performing sabotage:

  - **SHOTGUN**—See above.

  - **HIDING**—In this instance, don't hide yourself, but the other traveler's time machine. Seafloors and crevasses are preferred to under bushes, behind billboards, inside barns and other standard time machine hiding places.

  - **STEALING**—The problem here is that if you steal another time machine, you will typically have to leave your own behind, unless one time machine can fit inside the other, which is unlikely and cumbersome to accomplish (unless you are fortunate enough to have a booth-based time machine). This is known as the Pregnant Time Machine Maneuver. ***Note:*** Be

---

121 Unless you have to. See previous footnote.

122 QUAN+UM does not advocate the use of murder as a means of solving problems or disagreements. Use your words.

sure the inner time machine has had its power source removed. Operating a time machine with a time machine inside it is to the space-time continuum as eating a cheese-rolled, deep-fried Twinkie is to your digestive tract. **Recommended:** If your enemy's time machine is better or has futuristic-looking ground effects.[123]

■ **BREAKING**—Nothing says "gotcha" like a quick escape through a wormhole, followed by your timeponent attempting to give chase in his or her own time machine, only to find it inoperable and perhaps accompanied by a clever and carefully placed note.[124]

> **POTATOES:** When inserted into tailpipe, renders once-working car-based time machines into useless radioactive port-o-potties.

> **WRENCHES:** Useful for beating flashy or beepy components.

> **SCREWDRIVERS:** Useful for gaining access to flashy or beepy components.

> **POWER SOURCE:** Disruption of power source is the most effective way to sabotage a time machine.

Note 1: Do NOT use shotgun, banana, wrench, or bare knuckles on power source.

Note 2: Simple removal of power source will suffice.

---

123 Ground effects are not actually used in the future.

124 Hard to go wrong with "See you next *time*," "You're it!" or "Cat got your flux capacitor?"

Note 3: Simple removal of power source often is not possible.

Note 4: Do not remove power source with bare hands. Radioactivity answers to no clock.[125]

Note 5: Do not store removed power source near loins.[126]

TIMEBATTLING YOUR TIME SELF

Far from fighting off a hostile time traveler, this represents your worst—and also best—case scenario. Note that any time you run into a version of yourself from before you were a time traveler, yourself (if you recognize yourself) will undoubtedly try to kill you. It's called the Doppelgänger Effect, and it's a natural part of human evolution.

But if yourself knows about time travel and is aware of the eventuality of seeing yourself while traveling through time, yourself should be intelligent enough to be able to countermand the evolutionary imperative to murder yourself. But beware! Yourself is still dangerous, and since time travel is all about correcting mistakes and preventing things from happening, you can be sure that yourself will try to stop you from doing whatever it is you're doing or stop you from eventually doing whatever it is yourself is doing.

Inevitably, you will have to do battle with yourself. This will likely take place atop a cliff, during a lightning storm without rain. It may also include swords but is more likely to be hand-to-hand. You'll never have a gun-battle with yourself, so don't worry about that.[127]

---

125 With the notable exception of half-lives and carbon-14 dating.

126 Radioactivity is particularly devastating to the loins, as well as their fruits.

127 It can't really be considered a battle when you get shot in the face and it's over. There can be only One.

Phil Hornshaw & Nick Hurwitch

You do have to worry about yourself, however. You know all about yourself—your thoughts, your moves, your unsightly moles, your tickle phobia. You are the most dangerous opponent you will ever face, because you can anticipate you in every way. You do have one advantage over yourself, however—this guide.[128] And we're about to teach you how to beat yourself until you're exhausted, but fully satisfied, and thoroughly victorious.

But first you're going to have to gather some information.

"Knowledge is power."

—COLONEL SANDERS

ARE YOU TIMEBATTLING YOURSELF FROM THE PAST OR FROM THE FUTURE?

Are you fighting football star eighteen-year-old you,[129] or seventy-five-year-old emphysema-sufferer you? That is a key distinction, and assessing your physical abilities in relationship to yourself is knowledge that will help you gain the upper hand.

IS IT YOU FROM YOUR PAST OR YOUR FUTURE?

What does the you you're about to eviscerate know? Does he or she know about this guide and remember the information herein? Did your future self have bionic upgrades implanted to the ass-kicking tendons of his or her future feet, of which your ass is as yet unaware? This, too, is essential information. Yourself thinks he or she knows what to expect because he or she is you, but if you know more about yourself than yourself knows about you, then yourself won't know what to expect from you. This may be the only edge you'll have, but likewise the only one you'll need.

---

128   If yourself also has this guide, especially the same version of this guide from the past or the future with your handwriting in the margins, destroy that guide at once. For one, a book is much easier to kill than a human. For two, that is technically unauthorized duplication and our robot attorneys are chomping at the recently lubricated bit, as they were programmed to.

129   We're kidding, of course – by "football star eighteen-year-old you" we meant "Dungeons & Dragons enthusiast eighteen-year-old you."

Good—by making an assessment, you know what to expect from your fight, and you know what information and knowledge your opponent has at his or her disposal as well. Now it's time to put that information into action to ensure you beat the snot out of yourself:

AVOID PREDICTABILITY

Yourself, more than likely, fights a lot like you, so you don't want to find yourself in an endless wax-on, wax-off mirror dance[130] in which neither of you can land a blow on the other. You need to anticipate what you're going to do and how to outsmart yourself. That means, in all likelihood, fighting like an escaped convict with schizophrenia and one bad eye. Get desperate. Get mean. If you tend to favor your right side, hammer away at yourself on that weaker side. If you know you have a tendency to flinch a lot, make a lot of loud noises and wave your hands in front of your eyes, then kick yourself right in your balls/ovaries/cybernetic gene propagation unit. We're not talking about fighting fair, we're talking about defeating a Hostile Time Traveler. This could be key to the survival of you and everyone you know.

---

130 Or worse: a whacks-on, whacks-off mirror dance.

USE TIME TRAVEL TO YOUR ADVANTAGE

Even if you don't know everything about yourself, you know a lot about yourself: Are you the younger or elder you in the fight? Young people are fast and stupid; old people are slow, frail and wise. That means you can adjust your fighting style accordingly: If you're the younger you, then use your speed and stupidity to your advantage, taking into account that yourself thinks you know yourself, so he or she will be expecting you to fight like yourself. So don't fight like you. Take advice from famed Hostile Time Traveler Defeater Muhammad Robotic Ali—float like a butterfly, shatter the hip. And if you're the older you, expect yourself to overestimate him- or herself and underestimate you. Fight with brains and experience, not power and speed. Step out of the way. Stick your foot out like a wise old kung-fu master and let yourself trip on it. Admonish yourself for getting that stupid tattoo, and distract yourself by showing yourself how wrinkled and gross it has become. Then kick yourself in your balls/ovaries/primary download extension port, because hell, you've already had kids and didn't like them very much any way.

PSYCHE YOURSELF OUT

If you're fighting you from the past, you have the ability to shatter your mind with accurate (or fabricated) predictions of yourself's future. If you're fighting you from the future, you know things about yourself—regrets, crushing defeats, childhood embarrassments, deforming injuries—that perhaps you haven't thought of in years. Psychology is your friend and a great tool: Don't keep your mouth shut, talk your ear off. Be as annoying, manic and depressing as possible. Tell yourself the things that would make you cry, then kick yourself in the balls/ovaries/cloacae.[131]

---

131   Note: Avoid subject material you have not yet worked through in therapy. If you also start crying, this tactic is kind of a wash.

HIGH RISK/HIGH REWARD COROLLARY

If you don't "finish the job" on yourself and leave an enemy or duplicate-and-hostile self stranded in some other time critically wounded or shanghaied after you've sabotaged his or her time transportation module, you will have achieved one of two distinct results (but very likely both):

RESULT 1: Victory!

RESULT 2: An über pissed off enemy who bides his or her time and dedicates the rest of his or her stranded life to reconstructing a time machine and ending your dirty time-battling ways, sort of as previously intended, but bolstered by anger, a lifetime of focus, brooding, and careful, vengeance-fueled planning.

The good news is that you will have your result faster than that of a self-aware robot pregnancy test. When you return home or to the intersection of spacetime where you indicated during the heat of battle you would like to retire after all this nonsense is over, if nothing happens, you have achieved Result 1. This is preferred. Due to the abbreviated nature of time travel, however, if Result 2 has occurred, your time enemy will likely be waiting there for you: a moment having passed for you, as if this is a continuation of the same timebattle; a lifetime of agony having passed for yourself— older and hell-bent on stiff-arming you into a black hole.

If Result 2 occurs, repeat the above techniques, only better.

(POTENTIALLY) SURE VICTORY

The best way to end the endless running, traveling and slapping that can become a battle with another time traveler or yourself is to meet it at its source. The source in this instance is not the location of the battle itself or even the location of the battle's inception, but rather some time before your enemy ever traveled through time.

Somewhere in time, the incident exists—always during the timeponent's original timeline, typically during the timeponent's formative years, as coming-of-age stories about the elderly are less common.

Your opponent may only be a child and will likely have no idea why someone in strange clothing that smells of gunpowder, plutonium and urine is trying to kill him, or get him to ignore what is clearly a time machine in the living room, but do not falter—this kid is trying to kill you eventually.[132]

Note that this is NOT an endorsement of child murder, or even child maiming: More often than not, you can achieve your desired result simply by causing your opponent to associate time travel with some trauma. For example, dress up as a ghost and shout "Time travel time travel time travel and you'll die!" Or just walk up to your opponent in a park and state, "I'm you from your future and here's what happens when you time travel," before releasing your opponent's dog into the woods.

TIME GAMBLING

Time gambling is a common but dangerous method of time travel exploitation that occurs when a time traveler, armed with knowledge of the future, goes to the past and bets on surefire winners.

A sports almanac is a quick and easy way to accomplish this form of timesploitation. For that reason, time gambling is sometimes referred to as the Biff Sports Almanac Perk, after the first time traveler to ever successfully get rich using the technique.

---

132 If you had a chance to stop child Ben Linus, wouldn't you take it? You better, dammit—John Locke is far more interesting than Jack Shephard. We need a QUAN+UM agent to go ahead and fix that whole "Linus murders *Lost*'s most interesting character" situation.

If you are interested in time gambling, there are some things you should consider:

- You are a degenerate.
- You could be using time travel to, you know, help people, or broaden your view of history.
- Okay, if we can't appeal to your sense of decency, at least remember that it is unwise to bet large amounts on too many events, as you will arouse suspicion.
- Likewise, it is unwise to win all the time, as you will arouse suspicion.
- Arousing suspicion is a good way to get your knee-caps pounded into dust and made into Super Glue by casino security.
- Spending more than three days at a time in Las Vegas or a similar gambling-fueled adult playground can result in insomnia, prostitution, buffet gut and Super Glue powder kneecaps.

All this means that you should keep moving, keep your bets reasonable and pace your winnings. However, like all forms of gambling, time gambling is addictive. It is common for time gamblers to eventually use up viable bets in viable eras and slip up, losing everything—or at least their kneecaps.

Some time gamblers get so desperate that they begin to make up bets and go back in time to ensure the occurrence of their chosen outcome. Take, for example, Hermann Fezzleburg, who once bet a wealthy industrialist that he could discover the missing reptilian-human linkage somewhere in Nova Scotia, Canada. The bet was for $103.4 million (adjusted for inflation) and led Fezzleburg to load up his time machine with scotch and go back in time with the intent to impregnate a prehistoric Gila monster.

Phil Hornshaw & Nick Hurwitch

He failed.[133]

Time gambling is related to timebattling because if you time gamble for long enough, you will eventually encounter another time gambler, which inevitably leads to dangerous one-upmanship and, once again, hand-to-hand combat.

If you or someone you love, or even just kinda like, has a time gambling problem, contact QUAN+UM immediately. That is, immediately after wrestling the sports almanac from his or her hands and placing the bet that he or she otherwise would have made. Money in the bank.

AFTER VICTORY

THE THEOREM OF INFINITE POWER THROUGH INFINITE YOUS, DECODED

If you are not yet familiar with the similarities and linkages between time travel and Multiverse Theory, you have some studying to do. In order to stay sleek and carry-able while on the run from medieval knights and FBAs,[134] and also so as to appease the brevity desired by our editor's incurably short attention span,[135] we have to keep The Guide to a somewhat unreasonable word count. Thusly, we can't keep repeating the same important things over and over again, even if it means putting your personal safety in jeopardy.

That said: Congratulations! You have defeated your timeself in timebattle and stand victorious. You are the superior (though not

---

133　This was also the moment at which prehistoric Gila monsters got the taste for human genital flesh, which nearly resulted in the extinction of the species.

134　Floating Brain Aliens.

135　In her defense, she has more than 2 million followers on *Twit*, the 4.7-character-limit social networking spambot.

necessarily original) you. Now for the bad news: If that whole multiverse thing is really how it is, that was only one of infinite yous available for, and theoretically capable of, fighting you.

In their tireless research, time scientists have uncovered an ancient Chinese nursery rhyme known as Yî, or "The One" in English. The One spread the world over beginning in the early twenty-first century due to the vast population of China and vast popularity of Panda Express. In it, a man known as Evil Jet Li travels from universe to universe, defeating himself in battle. Each time he does so, the "power" of the defeated Non-Evil Jet Li is evenly dispersed among the remaining Jet Lis, Evil and Non-Evil alike. Eventually, only two remain: Super Evil Jet Li and unassuming Super Non-Evil Jet Li, who is equally powerful but crippled by his goody-two-shoes nature. We won't spoil the ending, but eventually One of them wins, and there is only ever One Jet Li in any universe from then on.

We're afraid this is total bullshit.

Not to discourage you from trying, because, as you have now experienced, defeating nega-you in battle feels pretty boss. And doing it over and over again, if nothing else, is an endurance test of the tallest order.[136]

But what the nursery rhyme gets wrong is that the number of universes in which to discover and defeat Non-Evil Jet Lis is not finite. There are in fact in-finite Non-Evil Jet Lis in in-finite universes. There are also, theoretically, in-finite Evil Jet Lis as well. Need we remind you:

$$\infty - 1 = \infty$$

Infinity minus one still equals infinity, because infinity is an idea and not a number.[137]

---

136 Never minding that it is technically Suigenocide.

137 Despite what those dorks over in the graduate math department keep insisting to us over e-mail. Honestly, screw those guys.

Phil Hornshaw & Nick Hurwitch

That's why it's called infinity. The recombination of "power" derived from the defeat of one of these infinite Jet Lis, though still possible, is technically impossible to prove or disprove for the same reason:

$$\infty\,(\infty - 1) = \infty^2 = \infty$$

The power of infinite Jet Lis multiplied by the power of infinite Jet Lis minus one, now-defeated Jet Li, is infinity squared, which, due to infinity being infinite and exponents not mattering at all, equals infinity.[138]

Even if the power were evenly redistributed, it is being done so among infinite yous, so there would be no net difference in total power per you.

Now, we hear you saying, or at least hear the sound of your brain throbbing as you think, "Well, hey. Multiverse is just a theory, and even if it's true, no way in hell am I going to risk deatomization and time travel through Quantum Foam. No more infinite mess, right?"

We're afraid this, also, is total bullshit.

As we pointed out in the opening of the chapter, from the first time you time travel, your life ceases to be a straight line. Instead of existing at one point—at the previously definable "present"— you now exist at infinite points. You can travel to any infinitesimal instant during your life and discover another you. You can defeat a you, travel .015 seconds into the past, and fight you again.

You are, as far as you know, the only time-traveling you, but the longer you meddle in the affairs of Time, the less likely this becomes. So now there are infinite stagnant yous and, increasingly, infinite time traveling yous popping up here and there throughout time, throughout The Universe. You will soon kill for the days that everything was as simple as a regurgitated ball of yarn. Literally.

---

138 Here come the graduate math dorks.

6

THE WORST CASETIME SPACE-TIME SCENARIO: FIXING THE TIMELINE

WARNING: THE FOLLOWING information explores how to deal with the catastrophic failure of everything. Failure of this degree may come after you've encountered hostile time travelers, failed to defeat them, then stood idly by as they visited the past and began killing grandfathers. Or, after you've accidentally left a cellular phone in humanity's Dark Ages, triggering a super-acceleration of knowledge that ends with a nuclear holocaust led by Napoleon Bonaparte and an army of mind-controlled red pandas. At this point, you've screwed up so royally that you're cursing your infernal time machine, shaking a fist in the air, and declaring that no one man, woman, or super-intelligent ape should wield so much power.

Well, hate to say we told you so, but we wrote a whole section about this very moment.[139] You can refer back to that later for verification—later, because you had better not be standing there thinking that you can go destroy that infernal time machine

---

139 This is why you were inevitably destroying the time machines. See Chapter 3.

now, before you've mopped up the enormous mess you've created. We like the timeline as it is, thank you, and we don't intend to be dominated by apes or the French just because you decided to start following the rules a tad too late.

You've got work to do, intrepid time traveler. Stop crying over torn spacetime and get to it.

YOU BREAK IT, YOU BUY IT

Whether by your hand or not, the timeline has been rent asunder, and only a time travel–capable individual such as yourself can repair it. And we can resolutely assure you that none of us possess a desire to roll up the sleeves of our shiny biodegradable onesies to hop through time putting out your temporal spot fires. Gather your plutonium and era-appropriate helmet: You're going it alone.

But not just yet: First, you'll need to know exactly where and when to start your timeline scrub. This is extremely important information to have—without it, you can very easily start making things 100 percent worse.

Here's the good news: Since you're a time traveler and you're aware there's been a change to the timeline, you already have all the information necessary to repair it, whether you realize it or not. Your memories—of the original timeline and of your pre–time travelin' days—will be your guide.[140]

Provided you're not being irradiated, enslaved, turned into a battery by robots, or eaten by former-human cannibals, it's time for you to perform a little detective work. This is the first phase of timeline repair:

STEP 1: Map out the original timeline in as much detail as you can muster while keeping in mind that this is all totally your fault.

---

140 Disregard if your original timeline is in fact an altered timeline that someone else broke/created and subsequently decided not to fix/destroy. See section in Chapter 4 "Paradox by Inaction."

From here, there are several other vital pieces of information you will need to collect before heading to the past to save the future:

STEP 2: Figure out what the hell changed.

At some point between your traveling to the past or the future and your returning to the present/future/past, everything changed—that's the premise off which we're working. Your time machine appeared, the portal closed, and everything was just . . . wrong.

Using the accurate and detailed mental portrait of the original timeline, you can now begin to take note of how the present state of affairs differs from it.

Perhaps cities only exist under water, your grandma has a reptile tail, or Al Gore is the president of Canada: No matter how big, how small, how cool, or how strange, if it falls outside the acceptable boundaries of Marty McFly's Close Enough Axiom, something crazy must have happened between departing whenever you just were and arriving whenever you are now.

Some other common alterations to look out for:

1. Neighborhood bully deserving of comeuppance now unfairly a wealthy and corrupt multigajillionaire.
2. Someone with no moral compass or common sense now president, king, prime minister, or dictator of your country/planet.
3. Presence of underground movement to overthrow said neighborhood bully/disagreeable ruler.[141]
4. Alternate understanding of colors (e.g., red means go, green means stop).
5. People of your time fully resigned to the fact that they're being dominated by some other, previously

---

141 Especially if said revolution includes common people from your original timeline, such as bakers or teachers, taking on heroic roles in the name of humanity.

subordinate race of animal, being, household appliance, or plant life.

6. Previously dead (or alive) friend (or relative) now alive (or dead).

7. Friend (or enemy) now has a goatee (or is clean-shaven) and is evil (or good).

8. Technology notably advanced or receded from original timeline.

9. Deranged sense of fashion.

Taking note of such changes will help you with Step 3:

STEP 3: Figure out when the hell it happened.

You now know where you are and where you've been. Having witnessed alterations to the timeline, you can start to determine where things went wrong—because you were there. In fact, you probably did it yourself. We're certainly blaming you.

Think back. What did you (or some past or future version of you) (or some hostile time traveler you were supposed to stop but were too sissy to) change, and when did you (they) change it?

This is the last vital bit of information. It is impossible to fix the timeline without returning to the original point at which things were changed, because any effects wrought after the initial change take place on the new, altered timeline. Because of this, you'll always need to travel back to the moment of divergence in order to repair the damage. It should be noted, however, that the so-called "original" timeline is now lost—the best you can hope for is to mitigate the damage to such a degree that this "new" timeline—which is in fact just a messed up version of the old timeline—resembles the original timeline as much as possible.

If you're not sure exactly what you changed, you can still return to the approximate time you (or someone else) changed it and make like a Hardy Boy. Your temporal incongruity should mean you were at least present or nearby when the timeline-altering event took place, which will help kick off your clue-hunting detectivulary.

Once you have identified the approximate timeline-altering event and its approximate "location" on the timeline, you may finally step into your time machine and continue down the path to redemption or failure.

STEP 4: Go back to when you were and fix it.

Don't expect this to be easy—just expect that you have to do it. Not only are you serving your own self-interest in repairing the timeline you helped mess up, but you are also serving our self-interests as well, in that we are interested in still having selves. It is your duty to repair time for the good of all the people you left twisting in the wind under the hegemony of some idiot with a name like Biff. That's not something you just hoverboard away from.

As you've probably deduced by this point, timeline repair has no hard-and-fast rules, although standard considerations of causality still apply. Always keep in mind the things that you can alter, and the things that you can't. Some other provisos to keep in mind:

1. **YOU CAN'T INTERFERE WITH YOURSELF.** That's right—you can't just tackle yourself and prevent an event from occurring, as that creates a paradox. Unless you're excited about figuring out exactly what the horrific atom-smashing consequences will likely be (you'll probably melt or create a paradox wedgie using the fabric underpants of space-time[142]), you have to allow your past self to go through all the events that led you to coming back to repair the timeline.

2. **YOU HAVE TO REPAIR THINGS *AFTER* THEY'VE BEEN BROKEN.** It follows from number one above that you have to actually allow Past You to screw things up and ~~travel Back to the Future~~™ return to a later

---

142 Which are not nearly as soft as cotton.

point in the timestream.[143] Whatever gets changed, you have to do something about it after the fact. Otherwise, your personal history will be distorted, you'll never go forward to a messed-up future and then you'll never return to the past to fix it. The very act of trying to prevent a paradox would cause one: not the intended result. Hopefully you didn't crush a really important bug or accidentally hit a future president with your car when he was six years old or something.

3. **GET REASONABLY CLOSE.** If you do happen to accidentally hit a future president with your car when he or she was six years old, well—you might be shit out of luck. But instead of moping about and waiting for spaghettification to take hold, try to do whatever that kid-president would have done, or maybe write a book about subjects that would have become that president's beliefs, in hopes that perhaps some other kid will read it, internalize the notions and use them to become president—resulting in a Close Enough scenario. The Universe seems to be at least a little forgiving in this regard: Give it the ol' college try. It's the least you can do as penance for getting away with vehicular homicide.

4. **ACCEPT YOUR NEW DIVERGENT REALITY.** We refer once again to Marty McFly's Close Enough Axiom. If you return from the past and your girlfriend/boyfriend is hotter (or just plain exists now), or purple has somehow replaced blond in the genetic coding for hair, you should consider it a victory. We're not saying give up. We're certainly not saying "let the Nazis take over" or "let humanity fall to the

---

143 We are using a non-copyrighted phrase. There is no need for legal repercussions.

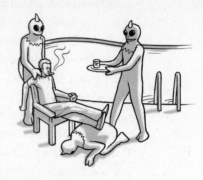

The Close Enough Axiom in Action

zombie apocalypse" or anything like that. But what we are saying is that you should be realistic about your expectations. Any future in which most of humanity doesn't consist of mutilated, irradiated, insane cannibals can be counted as a victory if you can achieve it over the alternative—a future in which most of humanity does consist of mutilated, irradiated, insane cannibals.[144]

5. **RETIRE FROM TIME TRAVEL.** Time is fickle and ruthlessly detailed: If you get through steps one to four, are "close enough" to your original timeline and are finding that no one notable is now dead or rising to alarming power, it is probably a good time to refer back to Chapter 3 and destroy your time module for good. It's safe to say that if you significantly altered history, we'd all be better off with you and time travel parting ways. Thanks for trying it out. Enjoy your nameless and boring life of obscurity.

---

144 And if your timeline-ruining alterations involved bringing back a once-dead relative, suck it up and see a shrink. The rest of us shouldn't have to suffer at the hand of Bat-winged Communist Cannibal Apes just because you are unwilling to deal with deep-seated regrets relating to your father.

Phil Hornshaw & Nick Hurwitch

Look, we know that up until now you've been taking the easy way out, but trust us—even Donnie Darko had to be convinced by the giant timerabbit to go back and fix the timeline that he also screwed up by leaving the timeline that the giant timerabbit made him leave in the first place. You don't want to be a giant timerabbit, do you? Actually, that's rhetorical. You don't want to be a giant timerabbit.[145]

It's going to take all your skills and brainpower, plus everything you learned in this guide, to save time. But you can do it. You have to do it. Help us, negligent time traveler in need of galactic redemption: You're our only hope.[146]

---

145 The authors are unsure of what, exactly, the hell happened in that movie.

146 Good. God.

TIMELINE

THE BEGINNING ~ 258 MILLION BCE

13,750,000,000 BC: The Big Bang.

1,000,000 BC: Bill and Ted stop in prehistoric San Dimas to fix their time machine.

986,473 BC: Rube Goldberg crash lands his time machine; stranded.

12,000 BC: The Doctor visits the Stone Age in the TARDIS.

258 MILLION BCE ~ 3300 BCE

1410 BC: Bill and Ted collect Socrates from Athens, Greece.

3300 BCE ~ 400 CE

0: Ya know. Jesus and stuff.

400 CE ~ 1300 CE

984: The last Dragon is slain.

1209: Bill and Ted snag Genghis Khan.

1300 CE ~ 1940 CE

1357: Ash arrives in England with only his boomstick, his chainsaw prosthesis and his 1973 Oldsmobile Delta 88.

c. 1450: 2003's Paul Walker and pals screw up the history of a battle between the English and French in Western France, as well as destroy any chance of a decent adaptation of Michael Crichton's novel.

c. 1450: Bill and Ted attempt to collect historical babes.

1492: Bill and Ted collect Joan of Arc from France.

1603: Four humanoid mutant turtles arrive in feudal Japan, which is convenient, because they are also ninjas.

1787: Alexander "Bear Arms" Hamilton disappears from his own time; Alexander "Bear Arms" Hamilton returns to his own time with a bowling trophy.

1805: Bill and Ted collect Napoleon Bonaparte in Austria.

1810: Bill and Ted collect Ludwig van Beethoven from Kassel, Germany.

1863: Future guys use 1994 weaponry to steal Confederate gold.

1863: Bill and Ted collect Abraham Lincoln.

1879: Bill and Ted collect Billy the Kid from New Mexico.

1885: A lightning strike sends Dr. Emmett Brown and his DeLorean time machine to Hill Valley's past, leaving from 1955.

1885: Martin McFly arrives in Hill Valley after having received the help of Dr. Emmett Brown in 1955, working to save 1985 Brown from being shot by "Mad Dog" Tannen.

1893: Jack the Ripper steals H.G. Wells' time machine and travels to San Francisco in the future. Wells gives pursuit.

1895: While riding a bus past a clock

tower, Einstein first imagines what would happen if he were to chase a beam of light and makes all this insane crap possible.

1899: H. G. Wells travels to the future. Again. Big damn deal.

1901: Bill and Ted collect Sigmund Freud from Vienna, Austria.

1919: Einstein marries his cousin.

1928: Seven years early, Delbridge Langdon III arrives in Poland, instead of Germany, with intent to assassinate Adolph Hitler.

1929: Jean Claude Van Damme arrives from 2004 in order to arrest his rogue Time Enforcement Commission partner, who's doing shit he shouldn't. In time.

1930: Captain James T. Kirk, Dr. Leonard "Bones" McCoy and Mr. Spock travel through time to the Great Depression.

1940 CE ~ 2040 CE

1954: John Locke meets Richard Alpert; Daniel Faraday checks out the Jughead nuclear bomb. Faraday doesn't realize he nearly creates a paradox by inaction because he's not as smart as he thinks he is.

1955: Emmett "Doc" Brown falls off his toilet and knocks himself unconscious. Using the power of brain damage, he invents the flux capacitor and makes a bunch more of this insane crap possible.

1955: Elder Biff Tannen from 2015

visits Young Biff Tannen and gives him the Sports Almanac. Sports betting is not a proper use of time travel.

1955: Marty McFly, having discovered Hill Valley all Biffed out from the use of the 2015 Sports Almanac, escapes back to 1955 with Emmett Brown's help in order to fix the original timeline.

1974: The survivors of Oceanic Flight 815 arrive from 2007 and join the Dharma Initiative. This probably occurs because of a naturally occurring wormhole.

1977: Three survivors of Oceanic Flight 815 jump back from 2007 when Ajira Airways Flight 316 crashes on The Island—more naturally occurring wormhole travel.

1979: Superman flies around the Earth, reversing its rotation, reversing time and saving Lois Lane with the dumbest plot device ever.

1979: H. G. Wells hunts down Jack the Ripper from 1893.

1984: Two sailors get zapped forward in time from 1943 following the Navy's attempt at making a warship invisible.

1984: Kyle Reese arrives from 2029 to protect Sarah Connor from a T-800 human-squashing robot.

1985: Marty McFly escapes Libyan terrorists by traveling back in time in the DeLorean Time Traversal Vehicle.

1985: Marty McFly and Emmett Brown return from 1955 to find that

TIMELINE

Biff Tannen has corrupted Hill Valley through the use of time travel.

1985: McFly uses the help of Dr. Emmett Brown in 1955 and a lightning strike to return to his present.

1986: Captain James T. Kirk and Spock go look for a whale to save the universe or something.

1988: Bill and Ted are visited by George Carlin in order to ensure a cool future.

1990: Time Travel Defense Magistrate Bruce Willis jumps back in time to stop a viral outbreak, but starts making movies instead.

1995: Dr. Sam Beckett becomes lost in time, possessing different people's bodies and talking to a hologram only he can see. Later, it becomes apparent he is just extremely fucking crazy.

1995: Skynet attempts to assassinate John Connor a second time by sending a T-1000 human-smashing robot after him. Arnold Schwarzenegger-bot intervenes on Connor's behalf.

1996: Time Travel Defense Magistrate Bruce Willis volunteers for a mission to go stop a renegade virus. Turns out it's SARS, and that wasn't a big deal because Bruce Willis saved our asses, guys.

2004: Two engineers invent closed-loop quantum displacement time travel and use it to game the stock market.

2005: The Doctor meets some random woman and brings her through time with him. Again.

2011: An evil quantum particle travels back in time to destroy evidence of itself at the Large Hadron Collider.

2012: QUAN+UM scientists arrive from the future to run first printing of *So You Created A Wormhole: The Time Traveler's Guide To Time Travel.* Critical reception is warm, but the Scientific community, composed of a bunch of meanies, largely views the book as a work of fiction.

2015: Emmett Brown, McFly and Jennifer arrive from 1985 to fix the McFly kids and nearly turn Earth into Biff's world, and implode the Universe.

2017: Delbridge Langdon III accidentally discovers a rare and dangerous form of time travel using static electricity.

2023: QUAN+UM founded.

2040 CE ~ 2183 CE

2043: Alexander "Bear Arms" Hamilton arrives at QUAN+UM headquarters with Ricky the Intern.

2054: A QUAN+UM research test trip to 1921 inadvertently returns with inventor Rube Goldberg, who decides to remain in 2054 and "look at all the neat stuff."

2056: Rube Goldberg travels to Prehistory during an unsanctioned QUAN+UM test trip; Goldberg returns from Prehistory babbling about some zany contraption. He is immediately booted back to 1921.

2100: The Starship Enterprise crew prevent the Borg from taking over the Earth, having arrived from the Twenty-Fourth Century.

2178: Man sent from post-apocalyptic underground Paris is gunned down on an airport tarmac, as witnessed by his five-year-old self. Successfully avoids Paradox by Inaction.

2183 CE ~ 2323 CE

2183: The Robots rise up against humanity.

2184: QUAN+UM headquarters destroyed. Survives underground.

2208: Scientists working in underground Paris send a man to the past using experimental dream-state technology in an attempt to prevent the robot uprising.

2286: Ash wakes up after taking a Medieval wizard's sleeping potion. Immediately realizes he has slept for too long.

2233: A Romulan ship travels back from 2362 through a Black Hole. We're not sure if we believe it, as it seems to screw up someone else's universe.

2258: Spock comes out of the black hole from 2362 for some reason.

2293: Humanity rises up against robot overlords; win freedom on a logic-based technicality.

2294: QUAN+UM headquarters rebuilt.

2313: Aliens invade Earth just as humanity re-establishes harmonious life above ground.

2323 CE ~ 13501 CE

2650: Humans and robots reach a pact with X-Filians, bringing an end to the long-fought war and occupation.

2673: Time Lord Doctor Who claims he gives booth-based time travel technology to George Carlin, who insists he invented himself, according to patent court documents.

2688: George Carlin departs to find Bill and Ted. Bill and Ted arrive in the Wyld Stallyns–inspired George Carlin future.

3978: George Taylor arrives at The Planet of the Apes, which is Earth. Taylor screams, "You maniacs! Damn you all to hell!"

13501 CE ~ ???

~4,730,000: The End of Time, probably.

802,701: H.G. Wells finds Morlocks.

THE TIME TRAVELER'S GUIDE TO TIME TRAVEL:

SURVIVING IN TIME

HOW TO USE THIS SECTION OF
THE GUIDE AS A FIELD MANUAL

AFTER ALL THE theory, methodology, machines and doppelgängers from the past and future, there comes a point in every time traveler's life when it's time to actually do it—to travel through time. But don't rush off into spacetime just yet, overexcited time traveler. For this will only be your folly. Stepping out of your time vehicle or applicable time portal will place you into the virtually unknown: You might think you know what dangers to expect at Woodstock 1969 from newscasts and Hollywood re-creations, but without bug spray, a week's worth of potable water and a powerful blunt object[147] for beating back stoned revelers, survival is questionable at best.

Here we've compiled basic information about each era of Earth's history to help you prepare for the worst case scenarios—information like how to fight off dinosaurs, how to turn a Great Pyramid into an impromptu Time Tomb Temporal Escape Device, how to survive the nuclear holocausts of the late twenty-first and twenty-second centuries, and how to determine whether you can make it with a hot alien babe.

This section of the book is to be used as a reference guide. Content is categorized first by era, then alphabetically except where otherwise noted. For example, if you're looking for tips on riding a dinosaur, you should look in PREHISTORY under "DINO-

---

147 Hit them with it quick, or the hippies will try and smoke it.

SAURS, Riding." Or, if you're unsure how to operate a microwave oven and are subsisting on frozen TV dinner rations, you should look in COMPUTER ERA under "ATOMIC AGE, Microwaves."

Also keep in mind that what lives in one era of time does not necessarily die in the next, or vice versa. For example, you'll find a handy section on Faking Swordsmanship in the EMPIRES section, but it's also useful for faking your way out of being murdered in the MEDIEVAL ERA. Running away from dinosaurs is covered in PREHISTORY, but the basic premise—"running the hell away"—applies to nearly any undesirable situation throughout history. Watch for footnotes referring you to useful information found in different eras of the Survival Guide.

SURVIVING IN TIME: UNIVERSAL RULES AND ADVICE

No matter when you're going or what you'll be doing there, time travel carries one incredible risk to the traveler: getting stranded in some crappy, dangerous era where you don't belong. As such, you should outfit your time vehicle or handy portal-ready mule with gear essential to your theoretical survival. The following are things you should always have with you and problems for which you should always be prepared.

WHAT YOU SHOULD BRING

This guide

Obviously.

"Boomstick"

Also referred to as "God-like Wrath" in certain sections of early history and prehistory, your standard twelve-gauge pump-action shotgun is easy to procure at any hunting store and is so visually impressive and functionally powerful that it can be useful in almost any era. And yes, you'll have to intimidate primitive locals by shouting "This is my Boomstick!" and firing a "warning shot" every single time you step out of your time machine. It's always new to them.

Running shoes

Never, ever embark on a dangerous trip through time without bringing along a fresh, comfortable, broken-in pair of running shoes. You wouldn't believe how many situations you're going to get into in which your life will depend on your high-speed exit, stage whichever. Trust the guide—always have a good pair of shoes.

Backup time machine battery

There are many eras in time that don't have running water, central air, or the most key of key time travel ingredients: electricity. Batteries also for some reason have a tendency to catch aflame during rough wormhole exits, so make sure you always have a spare charge at the ready.

Backup time machine

If you're bringing along backups, you might as well bring the one you really need. If you can afford a redundant time machine and fit it within your main time machine (even with all of the extra luggage you brought to put your souvenirs in), you definitely should. Time machines, finicky as they are, tend to only break when used. If you can use/break one to get where you're going, then use/break the other one to return home, at least you're home now, with no means or need to time travel again.[148]

Time machine components

Specifically we mean extra wire. QUAN+UM administrators cannot stress enough how useful it is to have spare conductive material to help you gather electricity and charge your battery.

---

148 Recall the Pregnant Time Machine Maneuver discussed under "Sabotage" in the Chapter 5 section "Timebattling Hostile Time Travelers."

Phil Hornshaw & Nick Hurwitch

Lots of things will (supposedly) give off power, but getting that power into your time machine is a job for wire. Bring that and any other components you can fit, especially knobs: Those plastic ones snap off all the time.

Potato seeds

You're planning for the worst-case scenario, and that is the scenario where you're trapped in some terrible era and have to live there for the rest of your life. Or at least until you can repair your time machine. That's where the mighty potato comes in: This energy-packed super veggie is easy to grow, can feed you when you're hungry and makes for an excellent small-scale battery capable of powering a lightbulb. Grow enough of them and you'll at least always have a delicious side dish as you struggle to cope with your ever-shrinking life expectancy. By the way, you should learn to farm.

FIGURING OUT *WHEN* THE HELL YOU ARE

Bad news, time traveling friend: Being trapped in time is often a function of accidental travel, or becoming lost. That means you may not always know in what period of Earth's history you've landed, and without that information, you're really up the river without a flat-headed composite titanium manual liquid propulsion rod. You're also going to have a tough time getting any useful information out of this guide.

Your powers of observation are your greatest tools for determining when you are. Look for the big indicators: language, technology and by what means the natives seem to be dying (other than from you).

The opening section of each chapter in this guide includes some important highlights of the era. Cross-reference these points to confirm what era you're in and also how screwed you are.

FIGURING OUT *WHERE* THE HELL YOU ARE

See above. This is just as important—You may be in 100 AD, but are you about to die of thirst in the Egyptian Desert or of abdominal lacerations in the Roman Coliseum? These, as you might imagine, are very different situations, even if they are in the same time.

Again, follow the points we've highlighted and observe before engaging whenever possible. Seeing a Japanese guy doesn't necessarily put you in Tokyo—you could be in Los Angeles. But if he has a sword, it's probably Japan (or a different part of L.A.). Use your common sense and your best judgment, but always exercise caution. Good luck, intrepid time traveler, and try not to get caught in any photos while holding nonexistent or vastly outmoded technology.

SURVIVING IN TIME: PREHISTORY

THE BEGINNING–258 MILLION B.C.E.

(Carnivorous Dinosaurs to Carnivorous Ice Age Land Mammals)

EDITOR'S NOTE: THIS section of the Survival Guide mostly deals in dinosaurs, but all solutions can easily be applied to various varieties of saber-toothed animals and other predators found on the Ice Age tundra. At least no one's told us otherwise.[149]

HOW TO TELL IF YOU'RE IN PREHISTORY
- Tar pits
- Large plants
- Large bugs
- Large animals
- Lack of humans
- Thundercats
- Pangaea

---

149 If you can handle a T-Rex, you can handle a giant cat.

- This guide
- Boomstick
- Running shoes
- Large-to-Xtra-Large flyswatter
- Backup time machine battery
- Backup time machine

INTRODUCTION

Though the term "Prehistory" technically applies to any time prior to recorded human history, for the purposes of this section we consider it to be the time that begins with dinosaurs and ends prior to the Dawn of Man. This is not because we are lazy, although internal performance evaluations have shown this to be the case, but also because if you go too far back in time, enumerating all the things and ways in which you could die round up to a simple "everything." The defining characteristics of Pre-Prehistory:

- No atmosphere
- No oxygen
- No ground
- Important early evolutionary events that we don't care to see because we don't care to screw them up
- Abundance of liquid hot magma

So if you find yourself having gone so far back that there aren't even enormous reptiles or razor-toothed birds that can swallow you in a single gulp, then there are no "tips" we can provide you other than "Hold your breath and jump back to the future as quickly as possible."

But if you see dinosaurs, well: Hold your breath and jump back to the future as quickly as possible. They are creatures that

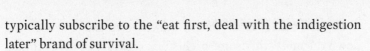

typically subscribe to the "eat first, deal with the indigestion later" brand of survival.

So welcome to Prehistory: where the plants are large, the animals are larger, the continents are singular, and any of these things, if given the opportunity, will eat you.

DINOSAURS

DINOSAUR FIGHTING

It's going to come up. Any time spent outside the time vehicle increases the likelihood that some large, hungry beast will want to pick a fight. Some things to try in a battle to the death with the Lizard King and his Lizard Disciples:

Boomstick

When you absolutely need to convince a dinosaur you'd give it indigestion, there's no better way to send a message. Unless that message accidentally ends up being "Oops, didn't mean to fire that" and you're now being chased up the steep incline of a volcano.[150]

Ammunition

You should have plenty, at all times. Strapped all over your body. Load up bags of the stuff. Time travelers have survived all sorts of situations with limited amounts of food, water and shelter—but the authors have never run across a prehistoric traveler who ran out of shotgun shells and lived to sign a book deal.[151]

---

150 See section "SURVIVAL GUIDE: PREHISTORY: RUNNING THE HELL AWAY."

151 Do not store ammunition or explosives in or around your time machine. Potential consequences of a misfire near flux capacitor include death, exploding death, death by singularity and death by ricochet.

EFFECTIVENESS: The standard-issue twelve-gauge won't bring down the most worrisome prehistoric creatures, but maximum effectiveness can be achieved by close proximity and due diligence to sensitive areas. Meaning if you can wait until you are about to be eaten and shoot the offending past-creature in the eye, or blast its mammoth[152] "jewels"[153] as it approaches, the shotgun is much more likely to save your life. Bonus tip: Multi-ton lizards are precariously balanced. A shot to a T-Rex's tail, for example, is worth about one hundred times more than one to its useless little carny hands.

INEFFECTIVENESS: Boomstick's effectiveness can be greatly reduced by the following conditions:

- Lack of ammunition (see above)
- A misfire (side effects: being eaten, stampede)
- Attempting to outrun anything not also carrying a shotgun
- Missing

No Boomstick?

- Look around: Can you distract it with a stick or bait? Do you have a slower, fatter comrade you wouldn't mind sacrificing?[154]
- Is there fire? Can you make a fire? You can travel through time, but you can't build a fire? Honestly?
- Adopt a boxing stance. Protect your face and body with your strong hand while jabbing at the dinosaur's

---

152 "Mammoth" meant here to serve as synonym for "large," "bulbous," or "humungo." Actual mammoth jewels are not to be shot; they have a hard enough time with boiling tar and developing glaciers.

153 We mean testicles. Did you miss the third grade or something?

154 If you are the slower, fatter comrade, make sure you are in sole possession of this guide.

eyes, nose and breadbox with your weak hand. When you've dazed it and it drops its guard, throw a haymaker with your strong hand. Remember that the strength of your hit actually comes from your legs—rotate into the punch. Right in the kisser.

- Can you construct a rudimentary firearm from anything that's lying around? Are there any stabby implements handy? Get to stabbing!
- Try reasoning with the dinosaur. Explain your situation and try to find some common ground. Know any jokes?

If all else fails, see "PREHISTORY: RUNNING THE HELL AWAY."

DINOSAUR RIDING

The preferred option to fighting. Though often you must fight (and defeat) a dinosaur anyway in order to do so.[155] This option also gains you some dino-clout (see also "DINOSAURS: Living with Them") and makes long-distance travel a breeze.[156] Here's how it's done:

- Assess available dinosaurs. You don't really want one of the ones that might like to eat you, unless you and the dinosaur have a previous understanding achieved through battle. Size is also a factor: Triceratopses are nice, because they're low to the ground and have horns for easy mounting; they are, however, stupefy-

---

155 Bonus points for outfitting your defeated dinosaur with a saddle made of the skins of other dinosaurs you've defeated.

156 Dismount at least a prehistoric mile (approximately 16.3 human miles) from your time machine to avoid having your time machine crushed, or accidentally transporting an apatosaurus into Cleveland.

ingly slow, spend most of their time eating and will charge headfirst—which means you-first—if threatened. Large, plant-eating dinos are a good choice, so long as you are adept at climbing impossibly high cliffs and trees for mounting, dismounting.

- Observe the dinosaur carefully: Does it have a weight problem or other psychological issue that you could exploit? Fat dinosaurs are more likely to let you ride them because they like the attention.
- Talk to the dinosaur. Pet its face and look deeply into its eyes. Whisper incoherent things to it. If it humps the air, licks you, or completely ignores you, you can probably just climb right on.
- Take a length of rope[157] and tie a loop.[158] While riding on horseback or some kind of moped, tear through the middle of a herd. When you spot an animal you feel you should be able to tame, spin the rope overhead for momentum and throw it around the dinosaur's neck. When it's too tired to continue dragging you, climb on its back and give it a name. We recommend Taxisaurus, or Clyde.
- Drop from something tall, such as a tree or cliff, onto your target dinosaur's back. If you can hold on for longer than eight seconds, the rules say the dinosaur has to let you ride it from then on.[159]

That's it! You're riding a dinosaur, or comparable ancient bi- or quadruped!

---

157 You brought rope, right? Did you bring *anything*?

158 For appropriate knots and how to tie them, refer to section *Handbook: Boy Scouts of America*.

159 Certain dinosaurs have tiny brains and, therefore, notoriously short memories. You may have to repeat this step once hourly.

Phil Hornshaw & Nick Hurwitch

USING THE DINOSAUR YOU'RE RIDING TO FIGHT OTHERS

Most dinosaurs aren't really interested in fighting you so much as stepping on you or biting you in half, seeing as you're not really much of a contender with something that's three stories tall unless you really brush up on your boxing skills. Instead, use your new dinosaur friend to fight your battles for you.

- Steer the dinosaur you're riding toward the dinosaur you want it to fight.
- Shout something like "Brachiosaur, I choose you!" so your dinosaur knows what's about to go down.
- Watch the dinosaurs' eyes to make sure your dinosaur isn't friends with the dinosaur you're about to fight. Otherwise you could be fighting two dinosaurs in a tag team–style cage match of the "two dinosaurs and one man enter, two dinosaurs leave satisfied" variety.
- Place bets with your bookie—at an unsanctioned event, throw cash down in the ring. As soon as there's contact, all betting is closed. Dem's the rules.

- Wait to see if your dinosaur turns and runs away from the other dinosaur. If so, jump down. See also "RUNNING THE HELL AWAY."
- If your dinosaur is committed to the fight, tell it you'll be waiting in its corner with water and gauze. Then jump down. See also "RUNNING THE HELL AWAY."
- If you lose: Say a small prayer (or comparable gesture appropriate to your belief system) in thanks for your dinosaur's sacrifice as you escape unhurt.
- If you win: Collect your winnings. Use the defeated dinosaur's skin as armor for your dinosaur so that it is more likely to defeat more dinosaurs and collect their skins as you make your way to the title bout.

LIVING WITH DINOSAURS

An ugly truth you may not have yet realized is that repairing your time machine in Prehistory, regardless of the amount of poop we recommend you've gathered,[160] may be (and more than likely is) impossible. If that's the case, settle in, because you'll need to make a life for yourself in your new time period.

LODGINGS

You'll want to live off the ground. The higher the better, in most cases. The big carnivores can reach the height of a three-story building or more, so to avoid them you'll want to be up higher than that. Higher. Keep climbing. Okay. Build a small toolshed where you can house your handmade wooden circular saw, mi-

---

160 See section "SURVIVAL GUIDE: PREHISTORY: TIME MACHINE REPAIR, Building Your Unfossilized Coprolite Battery."

ter saw, lathe, hammer and nails that you fashioned with the metals you mined (see section "PREHISTORY: TIME MACHINE REPAIR") and forged in the fires of the nearest volcano. Then wander about the jungle, randomly cutting down trees to use as lumber.[161] You'll need a pulley system to get your planks up to tree-house height and a ladder, so go ahead and build those, too. If you didn't bring rope, there are plenty of all-natural tying implements housed in the stomach cavity of one of your enormous neighbors. In fact, while we're on the subject—a hollowed-out dinosaur might make a decent cave dwelling,[162] provided you can tan the hide and ward off scavengers. If you're lazy or squeamish, see section "PREHISTORY: SHELTER."

FOOD

Get used to dinosaur meat. Managing to kill just one should feed you pretty much forever.[163] For vegetarians, watch the indigenous animals for clues about what might not kill you. Refer to section "PREHISTORY: FOOD CHAIN."

MAKE FRIENDS

As with indigenous peoples around the world, finding a harmony with nature is key to your survival. Hang out with the herbivores. Learn their customs. Try feeding some of the little guys and see if they would go for being pets, or at least let you drain their poison glands for your poison-tipped arrows. Anything not interested in living in harmony with you is better off dead or in another jungle that you don't own. Shoot to kill.

---

161 This is an era unfamiliar with industry: exploit it.

162 And you thought they smelled bad on the *outside*.

163 Get to work on your palm-leaf refrigerator. Use the same techniques employed in the fabrication of useful household items such as the coconut radio.

IF YOU SEE JESUS RIDING A DINOSAUR

Run over to him and start screaming. He should be able to perform a miracle and rescue you back to your time, right after he gets done turning all the world's "dragons" into fossils and hiding them under the ground.[164] Don't tell him about your time machine, though—it's blasphemous.[165]

FINDING SOMEONE LOST IN PREHISTORY

The good news is if you see a person in Prehistory, it's probably the person you're looking for. The bad news is, your maps are no good here, the accuracy of time travel decreases exponentially the farther you travel into the past, and your missing friend or loved one likely isn't sharp enough to survive in such an unforgiving environment or he or she never would have gotten trapped in The Land of the Lost to begin with. In any case, here are the steps for a successful rescue:

1. Look outside the time machine. Do you see the person you're searching for? If yes, proceed to Step 2. If no, proceed to Step 3.
2. Run out, grab that person, get back in the time machine and leave immediately. Congratulations on a successful rescue! The following steps don't really apply to you.
3. Give up. The person is dead.
4. Feel the initial, deadening shock of losing your beloved professor, time-traveling father, or soul mate.

---

164 Not to knock anyone's Creationist ideas or anything. Who knows—they could be right.

165 Just tell him you got drunk and woke up in 65,000,000 B.C.E., and you think your friends from the bachelor party might be responsible. If he needs convincing, draw a penis on your forehead before contact.

Phil Hornshaw & Nick Hurwitch

Deny to yourself that temporal relocation is even possible and that your superhero-esque loved one could have been tracked, hunted, attacked, incapacitated and eaten by indigenous creatures.

5. Acknowledge the pain of your loss. Cry. Harvest tears for potential drinking water (if stranded). Tell yourself if you had only been more of a genius, a faster time traveler, or a better student/son or daughter/soul mate, your lost comrade would still be alive.

6. Deal with personal guilt by releasing incredible rage on prehistoric life forms and time travel in general. Try not to fly off the handle and destroy your own machine, or kill anything that might somehow affect history. Attempt to bargain with God to save your lost loved one. Scream and yell in furious anger when God doesn't answer you— again. Try to come up with a plan that has you traveling to the past to stop your loved one from ending up a prehistoric entree.

7. Slip into depression as you realize you're probably not smart enough to save a dead person from a dinosaur in the past, regardless of your access to time travel implements and your childhood summer camp equestrian lessons. Reflect on the good times with the person and acknowledge your loneliness in this crazy, chaotic prehistoric world.

8. Work through your depression by cleaning the time machine as you decide that life can go on. Check all your dials and start thinking about the future. Or the past. Whichever.

9. Reconstruct the time machine or plan for your life without the person you lost. Realize it's not the dinosaurs' fault. Recognize you still have time-travely stuff to do, such as returning home, or building a tree house.

10. Accept that it's over, the dinosaurs won and time isn't waiting for you to get your act together. Start to feel hopeful that maybe this whole time travel thing wasn't in vain after all, and that you can do some good—like sell a dinosaur egg in the modern era for insane amounts of money.

FOOD CHAIN

You're not even on it. You're a smear on the snot guard at the buffet. You're slow and have terrible hearing, poor eyesight and almost no sense of smell. You make noise like a tone-deaf eighth grade marching band when you tromp through the jungle. Presume anything can eat you, even plants, and nothing is edible, even berries. That berry is millions of years old anyway, and if you're not poisoned just by looking at it, consider yourself lucky. Hope you brought PowerBars.

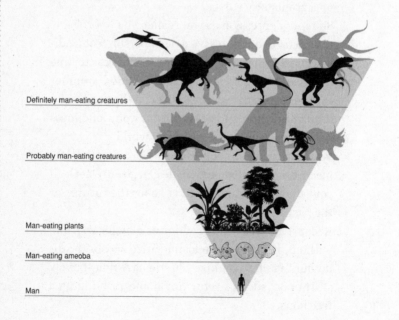

Definitely man-eating creatures

Probably man-eating creatures

Man-eating plants

Man-eating ameoba

Man

Phil Hornshaw & Nick Hurwitch

OCEAN

What the hell are you doing near the ocean? You shouldn't stay. Three reasons:

1. Loch Ness Monster
2. Greater whiter sharks
3. Dinocroc

See also "PREHISTORY: RUNNING THE HELL AWAY."

POOP

ADVANTAGES:

Electricity-producing microbes. See section "PREHISTORY: TIME MACHINE REPAIR, Building Your Unfossilized Coprolite Battery."

Hiding and camouflage. See below.

Animal feces are great for masking your mammal-stink, which means fewer terrible lizards drawn to you like fat kid to ice cream truck. Note, however, that not all poop is created equally. Choose wisely.

CARNIVORE POOP

While this brand will help scare off large dinosaurs of the plant-eating variety, such odors also signify territorial markings. The stench of a T-Rex, for example, could draw a defensive, emotionally crippled allosaurus that feels the need to display its masculinity through violence. And if it's mating season, which for all we know it always is, the aroma of female fecal matter could arouse the attention of a male dino in need.

HERBIVORE POOP

Less pungent and fibrous than its carnivorous counterpart, herbivore poop affords you a measure of invisibility to the majority of plant-eaters. They defecate with such volume and regularity that being covered in a thick blanket of diplodocus-dirt means you're about as interesting as the ground they walk on. Or, likewise, a pile of turd. But if you go this route, stick with the herb herd: There is safety in numbers. Alone and you are inviting the same sharp-toothed attention your nosedive into prehistoric dung was meant to prevent. There is such a thing as being eaten with dignity.

DRAWBACKS: FLIES. SEE BELOW.

Giant, terrifying ones. Bigger poop means bigger poop-loving bugs. If you're covered in poop, logic would follow that such a combination could go poorly, especially if the flies decide to eat your head, which is delightfully crunchy and easy to digest.

SHELTER

CAVES

Despite conventional wisdom, caves in Prehistory are occupied. All of them.[166] They're the Manhattan high-rises of pre-architecture. Think you're the first thing to ever want out of the rain/ice age/meteor apocalypse? Trust us: It's taken. It's also really, un-

---

166 Although consider fashioning one of your own from a delicious dinosaur carcass. See section "SURVIVAL GUIDE: PREHISTORY: DINOSAURS: Living with, Lodgings."

waveringly dark, and likely that anything within the darkness will eat you. Are you getting this? No caves.

PLANTS

Trees and large plants are also likely to be occupied by things that want to eat you. Under the trees lie bus-sized millipedes; in the branches rest large, ornery birds and flying dinosaurs protecting their eggs; and inside the tree—hello, that's still a cave: see above. Another consideration if sleeping in or around a prehistoric plant: It, too, could eat you.[167]

---

167 For more on trees as homes, see "SURVIVAL GUIDE: PREHISTORY: DINOSAURS, Living with," where we make the suggestion that you live in them despite the fact that they might eat you.

DON'T LEAVE YOUR TIME MACHINE:

Look out any windows your time machine may have and answer the following questions:

1. Is anything huge and salivating waiting to pounce on me?
2. Is my machine currently being ground between the teeth of something house-sized?
3. Did the arrival of my machine crush anything's eggs, nest, home, children, or the like?
4. Am I sinking in a tar pit?
5. Have I just spent the last five minutes staring into my time machine's toilet, thinking it was a window?

If you answered no to all these questions, then it might be relatively safe to go outside.[168] And in Prehistory, relatively safe is not safe at all. Your time machine is your home, your shelter and your way back to a more civilized era with slavery and war.

If your time machine appears inoperable, see section "PRE-HISTORY: TIME MACHINE REPAIR."

TIME MACHINE REPAIR

So long as your time machine is airtight,[169] regardless of what's wrong with it,[170] there's one quick fix: Blast it with electricity. Powering up your vehicle is really all that's necessary to open

---

168 If you answered yes to Question 4, you should leave your time machine immediately, regardless of other safety considerations. Please note: Gift receipts void if guide's pages become gooey with tar. This guide may not be used as a flotation device.

169 If your time machine is not airtight, maybe try sealing it with mud. If you don't cook from the inside, let us know. We'll put it in the next edition of this guide as a viable solution.

170 Only applies to actual time machines.

Phil Hornshaw & Nick Hurwitch

a wormhole to some time that includes sufficient materials for fixing it.

Electricity. Not exactly in abundant supply in Prehistory, the intrepid traveler is likely thinking. But the intrepid traveler would be wrong. Turns out, microbes found in animal intestinal tracts generate electricity. We researched it—it's Science.[171]

Intestinal microbes are found in—you guessed it—animal droppings. And what do you have in abundant supply? Dinosaur poop.[172]

BUILDING YOUR UNFOSSILIZED COPROLITE BATTERY

Here's the bad news: While it's possible to build a giant dinosaur-dung battery capable of powering a one-shot trip through time to escape Prehistory, you're going to need lots and lots and lots of poop. The amount of poop needed to generate the 1.21 gigawatts necessary to open your wormhole is, roughly, 600 million liters.[173]

Wire is also necessary. If you brought some, great. If not, mining is the next option. Good luck. Then, follow these easy steps:

1. Procure copper wire and steel from your time machine, as well as steel wire. You have that stuff, right? (A penny and a galvanized nail will work, too.)
2. Gather 600 million liters of droppings, preferably from the same species of animal. We'll wait.

---

171 Science based on cow droppings. Extrapolation to long-extinct reptile pies merely speculative.

172 Science prefers the word "coprolite." But coprolite is *technically* fossilized, so for our purposes we'll stick with poop.

173 This may take you as long as two years or more to obtain. Well, sorry, do you want to get home or not? Next time build a more reliable time vehicle. Besides, let's face it: You only have one thing better to do (see section "SURVIVAL GUIDE: PREHISTORY, DINOSAURS, Riding").

3. Pack the poop together tightly so it conducts electricity. Wash your hands before eating.
4. Stick the steel wire in one side of the poop (this is your cathode, the negative terminal of your battery).
5. Stick the copper wire in the other side of the poop (this is your anode, the positive terminal).
6. Connect wire from each end of the poop pile to your time machine's battery, but don't make the last connection until you're ready to go.
7. Make the last connection while standing in your time machine.[174]
8. Are you still standing beside a huge pile of dinosaur dung? Did it work?

IT DIDN'T WORK

Sorry.

Alternative:

1. Move the time machine into a big, wide-open field.
2. Retrofit as much metal as can be spared into a long pole.
3. Stand the pole straight up into the air.
4. Connect the pole to your flux capacitor or battery.
5. Hope like hell lightning strikes you.

With luck that gets you at least to some point in history that includes metallurgy.[175] If neither solution has worked, you might want to get comfortable, because it's Robinson Crusoe time.[176]

---

174 If you screw this up, your time machine goes, you stay, and we write a cautionary chapter about your horrible, what-not-to-do death.

175 Possible side effect of the Lightning Blast from God and Dino Dung Battery methods: being shot farther into the past. If this occurs, repeat all steps.

176 See section "SURVIVAL GUIDE: PREHISTORY: DINOSAURS, Living with Them."

RUNNING THE HELL AWAY

Abandon the time machine. Toss the shotgun. Forget the poop.[177] Your last, best option in every situation in Prehistory that includes encounters with the indigenous population is to just run.

Though be warned—you'll probably die anyway. But at least you'll die with an adrenaline high and a new personal best over whatever arbitrary distance you just covered (thanks to the running shoes you were billiantly advised to wear). You don't want your last thought—as teeth the size of bananas sink into your flesh, your own shattered bone fragments pierce into your muscles, and you lose the ability to differentiate internal from external bleeding—to be "Damn these stilettos!"[178]

177 See section "SURVIVAL GUIDE: PREHISTORY: POOP." See also "SURVIVAL GUIDE: PREHISTORY: TIME MACHINE REPAIR, Building Your Unfossilized Coprolite Battery."

178 Studies have shown people who have had near-death experiences feel much more fulfilled when their final thoughts are more along the lines of "Ow" and less "Man, am I an idiot. Ow."

If you don't fancy yourself much of a runner, get back to your time machine and refer to section "PREHISTORY: SHELTER, Don't Leave Your Time Machine."

SURVIVING IN TIME: DAWN OF MAN

258 MILLION B.C.E.–3300 B.C.E.

(Irritable, Club-wielding Cavemen to
Impatient, Spear-wielding Tribes)

HOW TO KNOW IF YOU'RE IN THE AGE OF EARLY MAN

- Tar pits
- Mammals
- Dead dinosaurs
- Several continents
- Cavepeople
- Ape people (pre-evolved)
- Lack of civilization
- Lack of chivalry
- Lack of Christ

WHAT YOU SHOULD BRING

- This guide
- Boomstick
- Running shoes
- Spare socks
- Sack of potatoes

- Impressive technology, such as a lighter
- Backup time machine battery
- Backup time machine

INTRODUCTION

There's a middle period between the dinosaurs and the rise of the durable upstart mammals that should be avoided. It mostly involves a giant meteor, fire in the skies, and a hundred years of suffocating, sun-free darkness. But once you're through all that, mankind is on an upswing. Tribal living, the invention of early tools and the first several lovemaking positions are all right around the corner. Luckily for you, fire is not yet en vogue, which, as you will see after a quick skim, is something you will need to use to your advantage in order to survive.

Though there is no longer the threat of giant warring reptiles in The Dawn of Man, the resources at your disposal and your chances of overcoming the wrath of untamed nature are about as disastrously low here as they are in Prehistory. Just keep your head up, your body hair thick (for warmth and social camouflage) and your time machine operable, and this whole terrifying affair will one day be a funny story you share with the kids over ice cream.

COMMUNICATION

Mankind's ancestors primarily communicate in three ways:

1. Grunting
2. Body language
3. Displays of dominance

These means of communication can be, and often are, intertwined. You will likely need to become master of all three if you wish to avoid being brain-pancaked in your sleep or having your skin torn off and worn as a shirt.

Curiously, it is widely believed that most Cavepeople from the Miocene period of the Cenozoic Era and on can, in fact, speak some form of broken English. Scientists have yet to determine a cause for this, but time travel is a leading suspect. One theory is that the producers of the short-lived 1960s sitcom *It's About Time*, unable to afford building their sets on outrageously over-priced studio lots, instead sent the entire cast and crew back to the Oligocene period to record the show on location. There, during union-sanctioned meal breaks, the crew's Bad English rubbed off on the surprisingly quick-witted Neanderthals.

This would, of course, mean that early language is in fact based on Bad English, and modern English is actually Bad English–based English, which would go a long way toward explaining why English was at once the most linguistically limited and culturally pervasive language on the planet during most of the twentieth and twenty-first centuries.[179]

Cavepeople may attempt to communicate with you using Bad English. Many will even have a superior vocabulary. But QUAN+UM strongly advises visitors of Early Man–populated eras to stick to the three native forms of communication listed above, as any further meddling in mankind's linguistic development could create big boo-boos and send you screaming back to Chapter 6—which would likely then be in some new, incomprehensible language and an uglier font, compounding your predicament a great deal.

---

179 The rule "'i' before 'e' except after 'c,'" for example, was actually one of the terms of an impromptu peace treaty between two tribes. There was also an exchange of large-breasted virgins as part of the deal, but that didn't really have any bearing on the creation of language.

GRUNTING

It's best to think about grunting not as its own language, but as an interaction with your dog or spouse. Your chances of grunting a combination of squeals and humphs that accidentally translate to "Please violently undress me and make with the Snoo Snoo" are rather slim (which isn't to say they won't make with the Snoo Snoo anyway). The Prehistoric Man's linguistic skills are not so refined.

Primarily what you are conveying with grunts is mood. Attitude. If your grunts are deep and fierce, the Cavepeople might fear you, or be less willing to share their mammoth meat. If your grunts are meek and high-pitched, the Cavemen may force you to walk on all fours and drink the breast milk of their females.

Just remember that instinctual responses are hardwired, even in the earlier iterations of humanity. Cavepeople deal with problems just like their genetically advanced brethren: with fear and violence. Be sure you use your grunts to offset the Cavepeople's collective lack of reasoned thinking and remorse to either calm, rile, or impress them, depending on your needs at the time.

BODY LANGUAGE

This is NOT to be confused with archaic forms of sign language. Things like pointing, waving, or other simple forms of physical communication mean very different things to a Caveperson, and the last thing you want is to leave an unprotected limb away from your body long enough to get smashed by a club, a typical Early Man reaction to sudden movements.

Body language is what your physical body says about you. Subtlety is not a strong suit of the Caveperson—you won't find any Napoleon-sized tribe leaders, or Cavewomen with equal voting rights. These are the days when the biggest men are the men in charge, and the smallest men are banished or used as toothpicks.

"I'M HERE TO STEAL YOUR WOMEN AND EAT YOUR YOUNG"   "PLEASE TAKE THIS ARM AS A SHOW OF MY WEAKNESS AND WEAR IT AS A TROPHY HAT"   "PLEASE TAKE THIS HEAD AS A SHOW OF MY WEAKNESS AND WEAR IT AS A TROPHY HAT"   "Hello! How are you this fine day?"

Caveman Interpretations of Common Sign Language

Thusly: Make yourself look big. Genetically speaking, mankind has grown taller over the years, but being tall is simply not good enough. Before stumbling into a Cavetribe, we recommend a daily regimen of raw eggs, protein shakes, and three months with a personal trainer. Or otherwise a complete abandonment of diet and exercise. Being repulsively obese is better than nothing; the Cavepeople might even think you are some kind of Pig God.

Body hair is also considered a symbol of status. Unless you have some kind of genetic deformity, or know how to brew Rogaine in your Hot Tub Time Machine, your paltry, modern degree of chest, back, leg and pit hair will not save you from being beaten to a pulp—no matter how envious or disgusted the other kids in the locker room once seemed.

A Bigfoot costume is a good double solution to this double problem. Just be sure no one touches your spongy skin or checks under your skirt for the other kind of body language. Yes, there.

DISPLAYS OF DOMINANCE

Hulking muscles, Rogaine baths and the right combination of gutturals should be enough to at least convince the Cavepeople to spare your genetically superior life. But to be accepted into a tribe and walk freely among them, something more brazen may be required—and we don't mean eating fist-sized fleas out of your neighbor's unwashed back hair. Though that, too, should be expected.

Other displays of dominance that may help or be required:[180]

- Killing an enemy Caveman in hand-to-hand combat
- Killing a tribe leader in hand-to-hand combat, thereby becoming the new tribe leader
- Wooing the most attractive Cavewoman[181] or Caveman
- Killing a mammoth on your own and offering it to the tribe
- Inventing farming (provided you have a little spare time)
- And last, but certainly first: the most compelling display of dominance this side of the Pleistocene period . . . Fire

FIRE

Fire is the life force of Early Man, the Boston cream to Center Camp's chocolate-covered glazed donut, and the main ingredient in Early Man diets, other than the actual main ingredients.[182] Making fire and making it well will net you free access to the tribes of your choosing, as well as myriad other social and survival benefits:

---

180 Panicked acts of panic that result in you asserting your dominance by showing off future pieces of technology (such as your time machine, or an Apple product) are a last-resort, panic-induced risk and should only be considered in times of panic. While this desperate act may temporarily distract or entice the Cavepeople, they will probably end up destroying the object, along with you, out of fear and jealousy.

181 See Section: "SURVIVAL GUIDE: DAWN OF MAN: CAN I MATE WITH IT?"

182 While fire may not be a food, Cavepeople aren't all that bright, and you may be able to convince them otherwise in a desperate moment.

Phil Hornshaw & Nick Hurwitch

- .Warmth—fire is hot
- Cooking meat—mammoth meat is tough and stringy when raw
- A light, for when times are dark[183]—sometimes it's dark
- Scaring off man-eating prehistoric cats or rival tribes—they are afraid and confused
- Alerting man-eating prehistoric cats or rival tribes as to your whereabouts—they are jealous and angry
- Deification—of yourself

This last one may be the most important: depending on how un-evolved the Cavepeople at hand, fire might not yet be invented. And even if it is, they probably aren't very good at making it yet.[184] How you choose to present fire to the Cavepeople is up to you—but should depend on how dire your situation.

If your personal safety is: unthreatened
Or your journey is: educational in nature
Then you should present fire: as knowledge

- The benefit of teaching Cavepeople how to make fire is that you will at once be revered for your knowledge and fearlessness in the face of something that really hurts when you touch it, and accepted as an equal by revealing that they, too, can burn things.
- The risk of teaching Cavepeople how to make fire is that, if they don't already have it (which is the only way your knowledgebomb will explode their walnut-sized minds), fire has not yet been invented.[185] You

---

183 Like at night, or inside a cave. Fire not guaranteed to aid in darkness that is internal or thematic in nature.

184 Fire log manufacturer Duraflame will not be founded until the mid-1970s, so it's very possible the Cavepeople you encounter will have no means of making fires of their own.

185 By "not yet invented" we of course mean that the means by which fire is

could set about an acceleration of human technology such that the first Pope of the Holy Roman Empire ends up with an iPod and the apocalypse happens full centuries earlier.[186]

- An additional risk is that once you show them how to make fire, the Cavepeople may soon discover how useless you are and decide to hurt your body and feelings, or banish you into the untamed world, where you are likely to die in fewer than three hours.

Remember that with this approach you actually have to know how to make fire. No lighters, propane tanks, flamethrowers, or burning this book: only things that can be found in Early Man's natural environment may be used.[187]

If your personal safety is: threatened
Or your journey is: panicked in nature
Then you should present fire: as magic

- The benefit of presenting fire to the Cavepeople as magic is that you will be seen as a god. You may even introduce to them the concept of god, thereby creating God in your likeness. This will be enough to buy you some time to hide, escape, or use your god-like powers

---

made and sustained by humans has not yet been invented. Fire itself, of course, was not an invention but a *discovery*—having for thousands of years mocked apekind with lightning-induced forest fires, lava-induced prairie fires and, eventually, spontaneous bush burnings. Mankind didn't start the fire, but it was always burning, more or less since the world's been turning.

186 Conversely, you should think back to Chapter 4 and the Paradox by Predestiny: Perhaps you were meant to teach the Cavepeople how to make a fire, and by not doing so, you could plunge humanity's evolution into a dark, downward spiral, resulting in a culture that reveres undereducated world leaders and waters its plants with Gatorade: *It's What Plants Crave!*®

187 If you use fire-making technology in the past, you could inadvertently drive some future anthropologist out of his mind when he discovers your two-million-year-old Zippo.

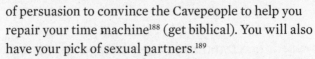

of persuasion to convince the Cavepeople to help you repair your time machine[188] (get biblical). You will also have your pick of sexual partners.[189]

- The risk of presenting fire to the Cavepeople as magic is that you will create a rift between you and them. You will never be accepted as one of them and your nearly interspecies erotic encounters will be hollow and unfulfilling. Moreover, as mentioned previously, Cavepeople tend to respond to fear and things beyond their comprehension with anger and violence. It will be only a matter of time before they steal your lighter or easy-start propane grill and, once failing to re-create the sacred flame, destroy it by smashing it against your face.

An Additional Warning About Fire Safety

FLAME IS HOT. PLEASE EXERCISE CAUTION WHEN STARTING FIRE. CONTAIN FIRE WITH A FIRE PIT OR RING OF ROCKS. COMPLETELY EXTINGUISH FIRE AND EMBERS WITH WATER OR URINE. DO NOT TOUCH BARE SKIN TO FLAME. FLAME IS HOT. IF YOU CATCH AFLAME, REMEMBER TO STOP CATCHING AFLAME, DROP TO THE GROUND, AND ROLL AROUND SO AS TO DISTRACT YOURSELF FROM THE SEARING PAIN OF THIRD DEGREE BURNS WITH ITCHY GRASS AND PUNCTURE WOUNDS. IF STOPPING, DROPPING AND ROLLING PROVE INEFFECTIVE, SMEAR BODY WITH MAMMOTH DUNG. IF MAMMOTH DUNG PROVES INEFFECTIVE, TRY WATER. ONLY YOU CAN PREVENT PREHISTORIC FOREST FIRES. PREHISTORIC FORESTS COVER WHOLE PLANET. IF YOU BURN DOWN WHOLE PLANET BEFORE

(continued)

---

188 See section "SURVIVAL GUIDE: DAWN OF MAN: TIME MACHINE REPAIR."

189 See section "SURVIVAL GUIDE: DAWN OF MAN: CAN I MATE WITH IT?"

DAWN OF CIVILIZATION, YOU ARE SOME KIND OF FIRE HAZARD SAVANT. PRIOR EXPERIENCE WITH NOT STARTING FOREST FIRES RECOMMENDED. FIRE SAFETY MERIT BADGE RECOMMENDED. WE ARE NOT YELLING. THIS IS JUST THE PROPER FORMATTING FOR SAFETY WARNINGS. HAVE YOU CONSIDERED THAT THIS IS THE ONLY WAY TO GET THE MESSAGE ACROSS? FLAME IS HOT.

SHELTER

Caves and trees are much safer dwellings at the Dawn of Man than in Prehistory—though there is still the odd flesh-eating creature or mammal with the natural instinct to disembowel anything that comes near its young.

Cave homes get the slight nod over tree dwellings. They come premade and provide better protection from inclement weather and wasp nests.

If you've made good with a tribe, or have been taken prisoner, chances are you're already living in a cave. If you're in a cave without a tribe, they're either on their way back or there is something wrong with that cave.

When choosing your shelter arrangements, it is important to consider your long-term plans.[190] If you intend to settle down and make a go of it at the Dawn of Man, you're going to need to consider a lot of factors about your cave and the area around it, including cave depth, underground monster hiding-place probability, nearby volcanic activity and proximity to a decent cliff off which to throw yourself to your death once you realize how terrible life at the Dawn of Man is.[191]

---

190 Of course, in this case by "your long-term plans," we mean "whether you crashed your time machine and marooned yourself in the belly button lint of history, you jackass."

191 Things that didn't exist at the Dawn of Man: FUCKING SOAP. Good luck sitting in an ever-present physically manifested cloud of your own failure and self-loathing.

Can I Mate With It?

Tight bodies, stunning athleticism, wildly unkempt hairstyles, the ability to nearly walk upright and a penchant for performing day-to-day tasks in the nude—though many are not even technically Homo sapiens, Cavepeople are far enough from ape-like that you may find yourself sexually enticed. Similarly, they are close enough to ape-like that you may find yourself sexually enticed. If you are still reading, having not yet slammed the book closed in disgust, you are likely wondering one of two things:

- Is that even possible?
- Is it, ya know . . . safe?

Nebulous questions, both. The nebulous answers to which, of course, depend on your definitions of "possible" and "safe."

If by "possible" you mean, "Is the Caveman or Cavewoman of my choosing anatomically equipped to handle me?" the answer is yes. But if by "possible" you mean to tell us that the twinkle in your eye is some sort of three-quarters human/one-quarter ape Lou Ferrigno hybrid-baby, for the sake of all that is holy we hope not.

As for "safety," it is important to bear in mind that Cavepeople are stronger and more aggressive than even your most physical former partners: The Cavemen make love like Cavemen; the Cavewomen, like volleyball players. If you happen to be the aggressor and don't mind a little untamed body hair, it is unlikely you will be refused. However, Cavepeople were raised well before the Age of Consent and even slightly before the Age of Chivalry, so chances are slim you'll have a say-so if one of them suddenly finds itself enticed by your bosom.

In either case, it is unlikely that you will exit the ordeal unscathed. The good news is that Cavepeople typically allow their sexual partners to live. The bad news is that the scars are permanent.

Some factors to consider when choosing the cave, both temporary and permanent, for you:

- **SIZE**—You'll want enough space to stretch your legs and maybe eventually raise your Cavebabies, but anything too large means that something too large will want to live in your cave, too.
- **LOCATION**—Is there a good place to hide your time machine nearby? A source of fresh water? How are the schools? You should research the quality of life in the area before moving in.
- **PREVIOUS OWNERS**—Are they coming back? If they come back, and you're in their home, you should make with the fire tricks and self-deification in a hurry. Just don't scare them so much that they stab you with a spear. If they don't come back, ask yourself why they might have left. Are they nomadic? Is there no food left to hunt? Is this the poo cave, where they do their pooing?
- **DÉCOR**—It's tough to hang things in caves, but most Cavelords will allow you to paint on the walls, so long as you limit yourself to stick figures and rudimentary interpretations of nature and sun worship. You should also set up your living space far enough in from the cave mouth that you're protected from the elements and not easily seen from the outside, but close enough to the cave mouth that you don't get lost or stumble into a Balrog.
- **VIEW**—Does your cave have a view? You can't put a value on a good view.[192]
- **BONUS TIP**: Don't keep your time machine inside your cave. But in the event that you have to abandon the

---

192 Unless you're a time traveler trapped in the Dawn of Man just trying to get out of a monsoon or shake the scent of a tribe of cannibals. Oh, yeah. Sometimes there are cannibals. Oftentimes, a view will be the least of your worries, unless maybe it helps you see the cannibals coming.

cave, you will have to abandon your time machine. It's best to leave your time machine under a giant leaf or some place obvious, because obvious places are the last place IQ-deficient Cavemen will think to look.

TIME MACHINE REPAIR

It was once believed that there was no reasonable way to repair your time machine at the Dawn of Man. Mammoths, large as they are, produce feces much smaller than dinosaurs' in size and volume. Collecting that much Mammoth mook simply isn't feasible.

But one time travel volunteer—thought lost forever, momentarily mourned, and quickly forgotten as lunchtime rolled around—returned to the QUAN+UM facilities from this unforgiving era haggard, beaten, and noteworthy in stench—but alive.

He debriefed QUAN+UM on a complex chain of apparatuses he'd built and daisy-chained, which together allowed him to power his time machine and achieve requisite wormhole-traversing speed, without the need for radioactive materials, the steam engine, or anything feces-related.

Some have called his makeshift jumble of daisy-chained apparatuses brilliant . . . others, needlessly complex. But thus far, Time Traveler Guinea Pig Rube Goldberg's method is the only one known to have successfully returned a live time traveler from the Dawn of Man in a time machine once broken. We're not saying you could do it; we're not even saying we could do it. We're just saying that it has been done. At the very least, this is how Rube Goldberg did it.

The accompanying illustration and following steps will teach you how to build a Rube Goldberg Time Machine in case of a Dawn of Man emergency:

1. Endear yourself to the local Cavepopulace. Do this by self-deification if you have to—they'll probably listen better if they're afraid you could burst their heads with your godlike wrath.[193]

2. Capture a woolly mammoth, shag carpet mammoth, or any equivalent large-assed, stupid animal. Stupid is a key factor, as is relative ass-girth.[194] [195]

3. Locate a high cliff, preferably with a fall of greater than one hundred feet. The ideal location will also have a large, open field at the top of the cliff. You're going to need a fair amount of space.

4. With the help of the local Cavepopulace, you'll need to construct a few things. Use their knowledge of the world around them to find appropriate materials, and your rudimentary knowledge of pet toys to construct ten large, human-sized wooden hamster wheels.

5. Have you encountered any large, human-sized hamsters? No? Okay, just checking. That would have been convenient.

6. You're going to need some . . . nontraditional components. Remember that mammoth you ~~enslaved~~ befriended? Did he have any friends? Locate an additional mammoth. Get your Cavepopulace friends to help you either (a) run the mammoth off a cliff, (b) stab it until dead, or (c) hold it down while you cut off a few big pieces of its skin.

7. Give the remaining mammoth to the Cavepopulace as a sign of good faith. Use your deification perks to ask your worshippers for a "sacrifice." You're going to need feet. And spare socks.

---

193 Time travelers who are paying attention should note that the words "godlike wrath" in the context of this guide will always refer to the close-range blast of a twelve-gauge shotgun.

194 Use the following scale when determining relative ass-girth: One mammoth ass equals approximately three Jennifer Lopez asses.

195 Regarding note 194: For our twenty-third-century readers: Jennifer Lopez was a late-twentieth-century/early twenty-first-century singer and "actress," whose defining feature was her mesmerizing, scientifically baffling caboose. An alternative scale: One mammoth ass equals approximately sixteen thousand pre-Core Meltdown Earth pounds.

8. Gather up several of your followers' severed feet. Make sure to leave a few followers un-sacrificed to run in the hamster wheels, help with other labor, and bring you a frosty beverage when you require one.

9. Attach socks to the severed feet. You can just staple the cloth part to the soles.

10. Build a smaller wooden wheel for each larger hamster wheel and, using some wooden gears (that you will have to make), set the hamster wheels so that they spin the smaller wheels when pushed.

11. Attach the sock feet to the smaller wheels with the bottoms sticking out.

12. Take some mammoth hide (with hair still attached) and stretch it flat between two sticks. Push the hide close to the foot-wheel so the feet will rub on it as they pass by it, generating static electricity.

13. Jam your spare wires into each of the feet and coil them together at the center of the wheel. Run the wire to your time machine's battery.

14. Repeat the process with the other nine human-sized hamster wheels.

15. Enlist ten Cavepeople to run on the human-sized hamster wheels, spinning the foot-wheels and generating static electricity using the woolly mammoth hides. This will take a great deal of running, so convince them to help using fire, leftover mammoth meat, or godlike wrath.[196]

16. While your Cavepopulace friends are powering up your time machine for your eventual return trip, you'll need to find a way to get the machine to the required speed of eighty-eight miles per hour for portal traversal. Remember that mammoth you captured earlier? Go make sure your primitive friends haven't eaten him.

---

196 See footnote 193.

17. Capture another mammoth.

18. Impress upon your worshippers that eating your captured mammoth is very, very wrong. Sacred animals are not for eating, tell them. Explain what "sacred" means.

19. Construct a mammoth pen with an open end facing the edge of the cliff. Make sure the mammoth can't go anywhere except over the cliff[197] [198].

20. Create a stable harness for your mammoth.

21. Fashion ropes from vines, mammoth ligaments, and anything else strong you can find. Attach the ropes from the mammoth harness to your time machine.

22. Construct a pair of strong wooden posts capable of supporting the weight of your time machine with Y-shaped tops. Feed the ropes from the harness over these posts.

23. Attach lead wires from your battery, and at the far ends leave a small pile of dry kindling that will easily catch fire.

24. Run a rope over the future fire location and attach it to a piece of wood hung from another rope, which holds it in position to swing forward when released.

25. Cut three hundred large domino-shaped rocks using your Cavepopulace and their Stone Age tools. These "rocks" can also be domino-shaped hunks of wood.

26. Position the first domino to be hit by the swinging piece of wood when its securing rope is burned through.

---

197 This might require the additional steps of building tiger pens surrounding the mammoth pen and then capturing tigers. We'll leave that for you to figure out. Just make sure that mammoth doesn't get any bright ideas about running *through* your pen—he must fling himself off that cliff for this plan to "work."

198 Regarding note 197: The Authors do not necessarily endorse this plan or any other as "working."

27. Position the other 299 dominoes in sequence so that they are knocked down in succession.[199]

28. Position the final domino so that it falls and breaks a tight string, which is attached to a stationary, taught bow and arrow.

29. Kidnap a small tiger cub from a local tiger playground, day care facility, or bus stop. Raw mammoth meat or candy can be used as bait.

30. Hang the tiger cub from a tree using a rope, suspending it by a harness. Aim the arrow from the taught bow at the rope.

31. The hanging tiger cub will attract a mother tiger, so be sure to hang the tiger cub high enough that it can't be retrieved by the mother.

32. Hang the tiger cub in such a way that when the arrow cuts the rope, the cub falls into a catapult below. The cub's descent will draw the mother.

33. Run trip wires around the catapult so that when the mother tiger approaches the baby sitting in the catapult's basket, she will trip them.

33. Attach these trip lines to the catapult's firing mechanism.

34. Build a firing mechanism that will make the trip lines fire the catapult.[200]

35. Aim the catapult at your caged woolly mammoth's ample backside.

36. Position the catapult beside a boulder so that the boulder will be pushed forward by the discharged catapult's arm.

37. Set the boulder in a trench so that it will roll to the gate, which bars random tigers from accosting your woolly mammoth.

---

199 Science categorizes this by using the baffling term "domino effect."

200 Do we have to spell out everything? Stop whining about not being a mechanical engineer with a minor in carpentry and just do it.

38. Create a latch that can be disengaged by the boulder when it reaches the gate.

39. Allow the gate to fall open, allowing the mother tiger to run at full speed directly at the kidnapped tiger cub and the confused woolly mammoth, which is still trying to figure why a tiger cub just bounced off its ass.

40. Watch from your time machine as the woolly mammoth, in confused panic, runs off the cliff to evade the tiger, which is actually attempting to rescue its cub and couldn't really care less about the mammoth.

41. Using the weight of your time machine and its wind resistance, calculate the amount of time it will take your time machine to reach eighty-eight miles per hour while free-falling through space.

42. Brace yourself as the mammoth's falling weight rips the attached lines and, subsequently, yanks you and your time machine through space and off the cliff.

43. Wait until your machine is at eighty-eight miles per hour in freefall, and at precisely that moment, trigger your flux capacitor to create a wormhole.

44. Arrive back in your own time.[201]

---

201 Okay, admittedly, this whole thing sounds a little fishy. Mr. Goldberg, for all his apparent genius, was a cartoonist and studied mining in college—in 1904. We haven't had occasion to test this particular method (nor has anyone, probably due to readily accessible godhood at the Dawn of Man), and so we're not sure it really works. But if you're desperate enough that you would cut off the feet of several pre-evolved humans, it just might be worth a shot.

Phil Hornshaw & Nick Hurwitch

DID IT WORK?

It didn't work.

SORRY ABOUT THAT.

1. Cautiously open the door of your time machine (if it isn't destroyed and if you aren't destroyed) and climb out, careful to avoid the giant puddle of mammoth guts you just landed in.
2. Get comfortable at the Dawn of Man, living the life of a revered deity. Conserve your godlike wrath for those moments you truly need it.
3. Return the feet.

SURVIVING IN TIME: EMPIRES

3300 B.C.E.–400 C.E.

(Human Sacrifice in the Nude to
Warmongers in Togas)

HOW TO KNOW IF YOU'RE IN THE AGE OF EMPIRES

- Empires
- Warring
- Slave labor
- Huge, amazing, impractical monuments
- Less body hair
- Rudimentary seafaring
- Early, revealing clothing (often resembling bedsheets)
- Jewelry
- Emerging nation-states

WHAT YOU SHOULD BRING

- This guide
- Boomstick
- Running shoes
- A bedsheet
- Backup time machine battery
- Backup time machine

INTRODUCTION

Paganism, Roman rule, huge mythological monsters—welcome to the Age of Empires, in which civilizations marched across the known world in order to meet their neighbors, stab their neighbors, plant a flag in their neighbors' town, and convince their neighbors to help build huge monuments for no money, mostly with the bargaining chip of—you guessed it—additional stabbings.

There's a lot to be learned in this period of time, which spans the growth of city-states and baby nations from Sumatra to Mongolia and back again. As with any time, there are also a lot of hazards, but this is one of the first eras in which other humans will, more often than not, actively work to make your life difficult. Gold reigns supreme, democracy and the rule of law are still just conjectural mono-zygotic sperm, swimming around in the prepubescent testicles of society, and many a time traveler has been conscripted by force into the manual labor behind the Seven Wonders of the World, never to return.

This is an era in which the diseases can kill you, the animals can kill you, and the people will definitely kill you. Plan accordingly.

While there are, of course, many empires to choose from, each with its own rituals of human sacrifice and different methods for the disposal of feces from within the common home, for the sake of brevity we're focusing on a few of the bigger ones: Egyptian, Roman, Greek and Chinese, with sprinklings of Mayan and Aztec empires, as well as the occasional Cradle of Civilization vacation spot.

BLENDING IN

It is easy, but important, at this time to blend in with the general populace in your empire of choice or unfortunate luck. It is increasingly difficult to pass yourself off as a god just because you have access to gunpowder, electricity and soap. That means you're much better off just trying to be "one of the Centurions": pick a fake career that plays to your strengths and offers you some free time. This is especially true if you're moonlighting as a mad scientist from the future trying desperately to repair his busted time machine and return to an era with indoor plumbing.

SLAVE

The basest of classes, this one involves all manner of manual labor, sexual gratification (not for you), and a whole lot of other awful things determined by your slave driver/owner. To be avoided.

Common in: Egypt, Rome, Persia

Less common in: Mayan/Aztec, Greece, China

- Seek empires in which you don't look significantly different from the people living in that empire.
- Try to maintain favor with people in power to avoid their forcing you to become a slave.
- Keep your snarky comments about men wearing togas to yourself.
- If a slave, keep in mind potential access to metals and other elements handy for time machine repair. However, free time to implement said metals is scarce.

FARMER

A step up from being a slave, but still involves a high degree of manual labor. However, freedom is a plus.

Common in: China, Rome, Greece, Persia

Less common in: Egypt, Mayan/Aztec

- Handy when you intend to grow a large number of potatoes to power a repaired time machine.
- Requires some skills a time traveler should have, which

can help save you from being on "get sealed into the pyramid with the Pharaoh" detail.[202]

SOLDIER

If you're male and in need of a career during the Age of Empires, be aware that you'll likely be conscripted into military service if you manage to avoid slavery. In fact, there are only two criteria to being drafted: (1) Be capable of holding up a sword, and (2) Be a functional absorber of blows, stabs and fire[203] while other, more capable soldiers actually get the job done.

Common in: All empires

Uncommon in: Rural areas not yet overrun by empires

- Allows a degree of respect and power (by force) over other citizens of the Age of Empires. Which also helps avoid becoming a slave.
- Soldiers tend to get killed, which is bad.
- It's tough to get time to yourself when you're being forced to charge city walls and swap war stories with the other soldiers whenever there's not warring to be done.

ARTISAN/CRAFTSMAN

Somebody has to make stuff in the Age of Empires, and you know about making, on account of you made that time machine thing that got you into this crazy mess.[204] With a little understand-

---

202 Although this isn't always a negative. See section "SURVIVAL GUIDE: EMPIRES: TIME MACHINE REPAIR, Tomb Time Machines."

203 If you find yourself in front of bigger, burlier soldiers who seem to regard you with disdain, your role is probably to be less sword and shield and more human shield and shield.

204 If you didn't make it, try looking at your time machine and figuring out how it was made with deductive reasoning. If you're not sure how deductive reasoning works, ask Socrates.

ing of engineering, metallurgy, or any of a number of other skills, trades and crafts, you can easily pose as the guy who makes the swords, rather than the guy who ends up stabbed in the chest by swords.

Common in: All empires
Uncommon in: Places where it's cold[205]

- Perfect excuse for when you're caught working on your time machine.
- Utilizes some basic skills you probably have.
- Good way to earn some extra money.
- Requires skills you will need to learn anyway for when your time machine is inevitably broken/stolen/built into the Great Wall of China.

PRIEST

Slightly more befitting of a time traveler's scientific know-how and modern sensibilities, a priest of . . . whatever the local religion might be can often gain respect without doing much of anything except waxing holy and spouting words and phrases of presumable importance.

Common in: Rome
Uncommon in: None

- Allows for respect.
- Does not require manual labor, or much doing of any kind.
- Makes explaining things like your time machine a whole lot easier. Your time machine is: an altar, a throne, a statue of a deity, some thing you found, a religious artifact bestowed unto you from [the] God[s].
- Remember: Prior to Jesus Christ, Moses, Muhammad

---

205 Good luck even tying your shoelaces when there's a blizzard slowly killing your noncirculating phalanges.

and L. Ron Hubbard, religion was largely conjectural, story-based and polytheist. If challenged, simply make up a new god.[206]

- Could potentially involve highly specific rituals, marriages and other events that require intimate religious knowledge. Expect that priests found faking will likely end up as slaves.

POLITICIAN

Politics changes very little throughout history. In essence: men standing in rooms shouting about things while less-privileged citizens or slaves do all the work. The same was true during the Age of Empires, and so the person doing the shouting may as well be you: someone educated and (potentially) literate. Then again, that may not be you unless you're really up to speed with your Latin/Mayan/Old Chinese.[207] In a bind, show your wisdom and savvy by cutting a baby in half or something.[208]

Common in: Most empires

Uncommon in: Mayan

- When you get to run things, people don't enslave you or make you run catapults, ride horses, or hide from arrows under a shield.[209]
- Tendency to get stabbed, however, as this was still considered an acceptable form of debate.

---

206 Try not to make up Jesus, unless you're L. Ron Hubbard. Then you can make up whatever you want.

207 See section "SURVIVAL GUIDE: EMPIRES: KEY PHRASES."

208 You know, like that King Solomon guy did. That earned him the title of "the Wise." It's an upgrade from your nickname, "the Idiotically Trapped in the Past Without Knowing the Language." That's what people call you behind your back, anyway. Their words, not ours.

209 Unless you commit treachery or are found out for your mutinous ways. Then its straight to the slave line.

- Increased scrutiny of your activities (and possible assassination attempts) can hinder time machine repair activities.

WANDERING MONSTER-KILLING HERO

The best (and worst) career in the Age of Empires is uniquely handy for a wielder of such items as those all time travelers should have on hand when they arrive in a hostile era. With their enhanced knowledge of weaponry and history, compounded by the fact that they should be carrying a shotgun around with them in a time when most shields are made of wood, Wandering Monster-Killing Hero becomes an ideal occupation for most time travelers.

Common in: All

- Great for earning respect, mates and free food in the Age of Empires.
- Loved by everyone, you can generally do whatever you want (within reason) without being hassled.
- Rarely murdered or enslaved by people.
- It's easy to work on a time machine when your only responsibility is to go adventuring once in a while.
- May require imposing physique, excessive weight lifting.[210]

WARDROBE

As mankind first discovered the benefits of clothing, they didn't discover much of it. The first rule of fashion, long thought to be "Don't Tuck T-shirts into Blue Jeans," is, in fact, "Less Is More." Typically a bedsheet (silk or cotton) or a jock strap (silk or leather) is all that you'll require to blend in.

---

210 Which, in this era, typically means goat lifting.

This time period you'll need to look no further than the local bazaar or street market to find the season's most striking fashions. Gold accessories and lightweight plate mail are in. Color fabrics and leg-covering slacks are out.

If you're petite, a belt around your waist will help to hint at and accentuate your curves. Conversely, if you've caught a few too many seeded grapes in your mouth this season, simply drape your sheetlike toga over your shoulders to cover any saggy areas or rolls of unsightly fat. The undyed fabrics of the period are very slimming, but beware of gusty winds and children with low vantage points: In the Age of Empires—that's right!—we're letting it all hang out.

Just remember, fashion troops, if you're looking to blend in, we have just three words for you: class, class, class. And that's not to say you're required to have any. It is to say that all of your fashion choices should be based on what class of people you're forced to live with, or what class you wish to infiltrate:

If you're poor, anything clean will not do. And any form of accessory, gold or otherwise, will get you pyramid-stomped faster than if you inadvertently scuffed the open-toed sandal of General Caesar. If you're looking to appear rich, don't overdo it; nothing's quite as embarrassing as drawing rave reviews for your gaudy twenty-four-karat Caesar Crown replica when you secretly have no other notable items with which to dazzle in the coming days and weeks. Rotate!

Conversely, soldiers have very specific wardrobes that are worn day in and day out, so accept no substitutes: maim a real solider or make a visit to the hot springs to catch one while bathing (no peeking!). Be sure you leave no shoulder pad or belt buckle behind, and practice privately to look like you've held a sword before.[211]

---

211 See section "SURVIVAL GUIDE: EMPIRES: WAR, Pretend Swordsmanship."

KEY PHRASES TO USE IN KEY EMPIRES

If you are linguistically impaired, use these key phrases to slip through noteworthy empires while maintaining a low profile. This will, of course, require that you brought with you an iDon'tRespectYourCulture Apple Universal Translator. And even with a translator, you don't want to go on about things that might arouse suspicion—or worse—run out the batteries in your iDon't before you've located the proper place to drain the lizard. Stick to these phrases and these phrases only or you will, once again, regret ever opening your mouth:

EGYPTIAN

- "Don't leave your mourning family to pick up the pieces: plan your afterlife now."
  - Useful if posing as a tomb insurance salesman, or a tomb robber posing as a tomb insurance salesman, and for angering those over age thirty-five.
- "What's up with all the cats?"
  - In case you're wondering what's up with all the cats.
- "I don't like sand. It's coarse and rough and irritating, and it gets everywhere."
  - For expressing irritation with overabundance of sand; for arguments relating to the inferiority of the Prequel Trilogy.

GREEK

- "Is that Mount Olympus in your tunic or are you just happy to see me?"
  - Common Greek pickup line.

- "Curse the gods who are cruel and vengeful, but less so the gods who bring wine and harvest!"
  - "God dammit."
- "What is the proper pronunciation of 'gyro'?"
  - To avoid embarrassment when placing an order.

ROMAN

- "Not the face!"
  - For when they go for the face.
- "Eat spear!"
  - For when you go for the face.
- "The oligarchy has truly offset the presiding will of its citizenry, what with its zeal for the debauched and disregard for the healthy makeup of its tender politicking; henceforth we shall wear only shrunken leathermakes against our loins, for even when it is cold, we shall show our displeasure of the equivocators whom represent us with merry folly."
  - Useful in the event that politics come up.

CHINESE

- "Take me to General Tso!"
  - In case you're hungry.
- "Wait, so samurai and ninjas are both Japanese?"
  - For expressing disappointment in ancient Chinese culture.
- "Man, Buddha's really let himself go."
  - He used to be so thin.

MAYAN

- "It smells like pee over here."
  - In the event of a nearby pee smell.

- "Don't take me! Take him!"
  - Useful if someone tries to human sacrifice you.
- "In case it ever comes up, avoid anyone named 'Cortez.'"
  - This probably won't make a difference, but it's worth a shot.

HUMAN SACRIFICE

As humankind became more civilized, public sacrifices of innocent humans became more rare, giving way to public sacrifices of humans deemed naughty by their governments. But in the Age of Empires, there is little to distinguish those worthy of human sacrifice from those worth sparing.

For instance, sometimes the frail and meek are sacrificed for the drain they place on society. Other times, the fit and healthy are sacrificed as a show of appreciation to the god(s). Sometimes those enraptured by the power of human sacrifice are slaughtered willingly, offering themselves to their god(s) happily. And other times, those opposed to human sacrifice are slain for their insolence.

A time traveler's best bet is to stay somewhere in the middle: supportive, but not overly so, and in shape, but not physically of note.

However, this balance can be difficult to achieve if your skin tone doesn't match those in the world around you—your white or black or brown or differing Tan Devil-ness will likely already have you branded as a prime sacrificial candidate or god of your own.

If you suspect you might be in line for sacrifice:
- Run.[212]

---

212 See section "SURVIVAL GUIDE: PREHISTORY: RUNNING THE HELL AWAY."

- Try to postpone your own sacrifice by suggesting others for, or shoving others into, the ceremonial blade.[213]
- Convince them that you are a god by using "magic," such as a lighter or, in an emergency only, your time machine (which they will probably destroy out of fear).

If you are being treated as a god and suspect you will be sacrificed so that you may return to the Earth/Heaven/Spirit Realm:

- Run.[214]
- Try to postpone your sacrifice by holding festive parties that center on non-sacrificial activities, such as dancing, eating, or charades.
- Complain about the quality of the instruments to be used in your sacrifice and make veiled passive-aggressive remarks about the favor you showed the neighboring village since they didn't "skimp on the catering" or "use the same knife they cut the guts out of slaughtered livestock with."
- Start murdering those around you to show what a fearsome and ruthless god you can become when threatened with things that are burny or pointy in nature.[215]

---

213 See section "SURVIVAL GUIDE: EMPIRES: KEY PHRASES, Mayan."

214 See section "SURVIVAL GUIDE: PREHISTORY: RUNNING THE HELL AWAY."

215 See section "SURVIVAL GUIDE: UNIVERSAL RULES AND ADVICE: WHAT YOU SHOULD BRING, Boomstick."

MUMMIFICATION/MUMMIES

Of all the things that the civilizations of the Age of Empires have in common, a fascination with drying out and preserving the bodies of the dead is easily the most ookie. It may also be the one that has the biggest effect on you, the person desecrating their graves. If you plan to use a mummy's tomb as the means to launch yourself back to more appropriate time coordinates, you'll undoubtedly have to deal with a mummy's curse, which is among the worst curses of Antiquity. Despite being dead, these guys are persistent.

WHAT YOU NEED TO KNOW ABOUT MUMMIES:

- Mummies are dead, which would make them un-threatening under normal circumstances, but also magical, which makes them capable of walking around and extracting organs from the still living.
- Due to being dead, mummies have limited conver-sational and reasoning skills. Avoid discussions of politics and philosophy.
- Mummies have an unhealthy obsession with inter-nal organs, mostly stemming from the fact that they have none. This makes them inherently dis-trusting and spiteful of those lucky enough to re-tain organs of any kind. That means you.[216]
- Open flame is the enemy of mummies, as are hun-gry moths. Unfortunately, open flame is also the enemy of hungry moths.
- Though almost invincible physically, mummies' egos are not nearly as bulletproof. They are par-ticularly sensitive about their dried-out raisin

---

216 Cutting out one of your own kidneys might help, but the speed you'll lose from open-back surgery makes the procedure kind of a wash.

skin, unsightly gait, and the fact that they're pretty much forced to wear white after Labor Day. Don't insult the mummies.

- Like everyone else, mummies love gold. They're a particularly greedy lot and are skilled nonverbal negotiators. These negotiations typically include a metal hook, their adversary's nasal cavity, and a canopic jar.[217]
- While mummies are interested in brains, the desire to acquire them has no gastronomic connotations, even though they are technically undead.

WHEN FIGHTING MUMMIES:

- Mummies are mostly invincible, so guns and Boomsticks are not especially effective. Unless, that is, you can use them to remove limbs. Dismembering a mummy won't kill it, but it will seriously hamper its ability to negotiate.
- Note that detached mummy limbs can move independently of their mummy bodies. Blasting off mummy limbs may only end up creating more enemummies.
- Don't bother unwrapping them. Your brain, saturated with years of Scooby Doo reruns, might suggest that the unwrapping of a mummy is its weakness and that under the bandages might just be a local gardener or the head of the teacher's union. Really, it's just an angry, undead, desiccated naked guy.[218]
- Always reserve the right to a strategic retreat. Lack of moisture in mummies' bodies makes their

---

217 We suggest a new negotiation strategy, time traveler: Let the mummy win.

218 It's a matter of pain. If you're about to die, is that really the last (shriveled) thing you want to see?

movements stilted and slow, which means you can readily escape them. There's just one problem . . .

- . . . A mummy will never stop pursuing you, no matter what. As the magical undead, they also will never tire or give up. A car-based time machine or other vehicle can buy you time, but eventually you will have to face your enemummy.
- Fire, while a mummy's natural enemy, also isn't a guaranteed mummy fight winner. It takes a long time for a burning mummy to disintegrate.[219] You might just end up with flaming mummy hands strangling you, rather than the regular, non-flaming variety of strangulation.
- Mummies have the strength of 1,000 angry chimps, which have the strength of 4,700 water buffalo, which have the strength of 8,465,962 genetically modified gnats. Do not attempt hand-to-hand combat.
- Mummies are curse-driven. They never just attack willy-nilly—they have purpose, usually in the name of vengeance or the return of a stolen item. That generally means you stole something from the mummy, and you and anyone involved with the theft are targets. Turn this into an advantage by lining up the appropriate human shields.
- Return what you stole to disarm a mummy's murder-rage. Who steals from the dead? Your mother raised you better.

---

219 Unless you have a giant charcoal grill cover.

SLAVE LABOR

No matter what the era, slavery isn't funny or acceptable. Unless it's being used to build one of the (as of yet not Seven) Wonders of the World. Such an accomplishment will easily outlast any malnourished slave blabbing about human rights or threatening to unionize.

There is a simple rule in the Age of Empires: Either drive slaves or become one. And trust us, you do not want to do either. But given the choice, you especially do not want to become one. The hours are inhuman, the labor is unrewarding, and the pay is insulting, even by your standards.

Some tips to avoid becoming a slave:

1. Avoid the poor and the homeless. You're a time traveler, not goddamn Robin Hood.
2. Wash your clothes every now and then. So the saying goes: "He who looks like a slave, it doesn't matter what you look like once you're a slave."
3. Avoid treachery. This is a one-way ticket to slavetown, provided those you crossed with your treachery don't have any hungry animals or gods to appease.[220]
4. Avoid senseless acts of nobility. Yeah, that guard is picking on that defenseless child. Ask yourself if you've ever before cared about the well-being of a child sweating in the desert in 1700 B.C.E.. However, if your heart is too pure for you to stand idly by, at least make a show of it. The average citizens who witness your good deed will be inclined to rally around you in revolt, or break you out of your shackles whilst a diversion is created by a deus ex machina.

---

220 See section "SURVIVAL GUIDE: EMPIRES: HUMAN SACRIFICE."

Some tips on being a slave driver:

1. Start small. Boss those around you around.
2. Stay small: You don't want to be driving around a slave that can kick your ass unless that slave is first in shackles and you know how to use that whip.
3. Sidewise grin. Practice it, know it, love it. No good pharaoh, king, or emperor will take your slave driving skills seriously if you can't first crack a fiendish grin. Cackling is a bonus.
4. Punish all slaves equally, regardless of their slavesmanship. You're only helping the slaves if you create a martyr.

TIME MACHINE REPAIR

If there is one thing that the Age of Empires has, it's famed, godlike leaders. And if there's one thing that famed, godlike leaders do well, it's warring and dying young. Which are two things. But if there are two things that famed, godlike leaders do well, then the third thing is building absurdly taxing monoliths, World Wonders, and tombs in their own honor. To fix your time machine in the Age of Empires, that's exactly what you're going to have to do.

FIRST, CHOOSE ONE OF TWO PATHS:

1. Become a famed, godlike leader and have a World Wonder built in your honor.
2. Retrofit someone else's World Wonder with time travel apparatuses.

As you might have deduced, Option 2 is a good deal easier than Option 1, but both present a fair number of challenges.

May not be to scale

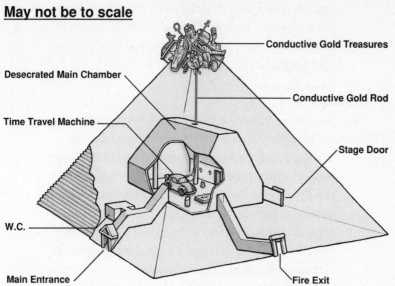

Conductive Gold Treasures

Desecrated Main Chamber

Conductive Gold Rod

Time Travel Machine

Stage Door

W.C.

Main Entrance

Fire Exit

Option 1

1. (a) Work your way into the ruling class by backstab-
bing politicians, sweet-talking other politicians, or
performing magic tricks for the emperor. Then,
when the time is right, claim the throne. (b) Murder the king/emperor/religious figurehead. Oftentimes this is all you will need to accomplish to
prove your worth. In the Mayan, Aztec, and Egyptian empires, he who kills the god-king must be a
more powerful god-king. In Rome, Greece, and
China, you'll at least have the support of those who
didn't support the guy you just murdered, which is
typically about half the populace.

2. (a) Take advantage of your popularity to have a
World Wonder built in your honor. Be creative. Or,
if you have replaced someone who commissioned a
World Wonder, just have that same one built so as
not to screw up history. (b) Have a World Wonder

built in your honor.[221] Refer to Chapter 3 or requisite time travel battery schematics (sold separately) in order to construct a large-scale battery out of your monolith.

Option 2

1. Distract or bribe the guards. This is easily done by walking past suggestively if you're female, or cross-dressing and walking past suggestively if you're male.[222] When they follow you around the corner of the pyramid, coliseum, or great wall, knock them out and take their clothes.
2. Now that you have free run of the World Wonder, you'll have all the time you need to fix the place up. If it's a tomb, you're in luck, because you can simply take the gold that's inside and hammer it to the top. Now, use the remaining gold to make a rod that runs down through the World Wonder and into your time machine.
3. Stuff your pants with excess treasure.
4. Wait for lightning to strike the gold, or . . . do a rain dance or something. Know any prophets in the mood to bring some plagues to your local empire? Call in some favors.

Bonus: Option 3

1. Become a slave.
2. Use your access to the construction of the World Wonder to implement timebattery elements throughout the building process.

---

221 Tombs are recommended because you will need to get both you and your time machine inside. And if things should go wrong, worst-case scenario, you're already buried.

222 For examples of how to execute this guard-distracting maneuver, please refer to the work of Chuck Jones, circa 1933–1945.

Phil Hornshaw & Nick Hurwitch

3. Choose between escaping in your time machine and freeing your fellow slaves. It is rare that both are possible.

WAR

You're going to get into some fights, and in this era you might not always want to go with "This is my Boomstick!" as your method for getting out of such situations.[223]

The only alternative available to you, however, is to use a sword (or some other kind of pointy implement). But let's be honest—you can barely handle that shotgun. Your weapons training consists of panicking at the sight of bees and flooring the gas pedal at small animals on residential streets. Apart from the influence of *Star Wars* on your adolescence, you have no idea how to handle a sword.

PRETEND SWORDSMANSHIP

There are two major reasons you can't allow yourself to look like a total buffoon with a sword in hand: (1) because you'll stick out as being in the wrong time era if you can't, since just about everybody can handle a sword in the Age of Empires, and (2) because you'll probably have to scare off some other idiot with a sword at some point, and if it looks like you're not sure which is the business end of the device, you'll fail to intimidate.

There are several steps you can take that don't require real training to at least appear as though you have some skill with a sword.

---

223 It's not exactly inconspicuous to go around exploding the faces of the people who disagree with you. It's generally frowned upon from the "don't go around murdering people" and "you really shouldn't be messing up the timeline with that shotgun all the time" perspectives.

① Stance

② Raise sword

③ Grip tightly

④ Hit opponent

⑤ Dodge opponent

⑥ Win swordfight

How to Win a Sword Fight

1. Throw out everything you remember from the third act of *The Empire Strikes Back*. It will not help you.
2. Stand with your feet spread comfortably apart and your dominant foot in front of you. Try to stand sideways so you present a smaller stabbing target.
3. Square your hips. You should have a nice, firm stance. Limpness is not impressive.[224]
4. Make sure you have a good, tight grip. You'll be moving fast and you wouldn't want the blade to slip from your hands like a sweaty lightning bolt, as this could result in injury.

---

224 If you have a limp stance, don't apologize, just fix it and come back when it's good and firm.

5. When the other guy gets close to you, try to hit him with the sword. It is not, contrary to popular belief, your goal to hit the opponent's sword. Finesse it a little bit—he's not a piñata. Swing too hard and you leave yourself open to counterattack. Not hard enough and that sword will come bouncing back at your face.

6. Do your best to not get stabbed, slashed, or otherwise injured. Use the skills you gained from your relentless dodgeball torment in school.[225]

7. Win the sword fight.

---

225 If you're from an era before the advent of dodgeball or after the banning of dodgeball by the Geneva Conventions, substitute the dodging your nerdy nine-year-old self learned from the applicable stoning/robot evasion experiences.

SURVIVING IN TIME: MIDDLE AGES

400 C.E.–1300 C.E.

(Religious Oppression—Black Plague and
Religious Oppression)

HOW TO KNOW IF YOU'RE IN THE MIDDLE AGES

- Jesus
- Inbred royalty
- Churches/stained glass windows
- Knights
- Crusades
- Stake burnings
- Development of the English language
- Quests
- Exploration

WHAT YOU SHOULD BRING

- This guide
- Boomstick
- Running shoes
- Sack of potatoes
- Bible
- Gunpowder/dynamite (or other magic-looking explosives)

- Backup time machine battery
- Backup time machine

INTRODUCTION

Things get a little hazy in the middle of ancient history, a section of time during which history was more diligently kept and yet was not quite populated with a whole lot of what we'd call "literate" members of society. Therefore, though there is a lot of "history" carried on from the medieval era, it is often wrong, exaggerated, or misremembered.

Here's a good example: dragons. Turns out—not a myth. That movie with Matthew McConaughey with the dragons and the end of the world missed out on box office riches by failing to include "Based on True Events" in its end credits.[226] At least, those are the accounts. As with many other eras in history, it's not like time travelers are just dropping in and out of periods like the Middle Ages. Lots go back and never return, for whatever reason. In this chapter, we'll try to prepare you to avoid being among their number. With that in mind, we're erring on the side of elves, wizards, dragons, and the flat world theory being real.

Should you find yourself trapped in the Middle Ages or planning to stay for extended periods, refer to the "CRUSADES" and "WIZARDING" sections for tips on blending in with people of the period.

THE BLACK DEATH

Fun facts about the Middle Ages: At one point, the population of Europe was reduced to almost half by a single disease, believed

---

226 Matthew McConaughey, however, is a highly "skilled" actor who fears no beast. You, on the other hand, should be much more afraid than you are.

to be carried by rat fleas and unsanitary conditions. Still sound like a fun place to visit?

Unfortunately, if your travels force you into the Middle Ages, it'll be difficult, if not impossible, to avoid such diseases,[227] or at least their aftermath.[228] Since gas masks and antibacterial soap are a long way off, in the twelfth century, you'll have to make do with a few other precautions:

AVOID POOP—We know—no-brainer, right? Apparently not. The Middle Ages were fraught with poop, along with garbage and a whole lot of other really gross things. There wasn't exactly curbside pickup, nor was there Ty-D-Bol or even the flush as we know it. Regardless, do your best to keep clear of unsanitary conditions, bring plenty of Purell, and watch where you step, for maximum disease-avoidance.

AVOID RATS—Modern rats might be furry friends you can carry around on your shoulder and teach tricks[229][230], but rats in the Middle Ages are carriers of disease and possibly murderous magic demons. They revel in all that garbage and poop you're trying to avoid. We're not saying "stomp rats to death if you see them"—in fact, you're going to want to flee, since it's actually the rats' fleas that carry bubonic plague.[231] It's hard to stomp a flea to death, and fleas have a

---

227 Outside of, you know, vaccines and medicine, to which you have access, being a time traveler.

228 Except by not arriving between 1348 and 1350. Then you can pretty well avoid it all.

229 Rats are really very smart.

230 This section sponsored by People and Rats for the Equal Treatment of People as Well as Rats.

231 We're reasonably sure Black Death was bubonic plague. Sure hope it wasn't, you know, something else for which we didn't inoculate any of our time travelers. Maybe we should have checked on that.

tendency to flee the body of something getting stomped to death for the body of the thing doing the stomping.

HOW TO KNOW IF YOU'VE CONTRACTED THE BLACK DEATH

Tally your answers to the questions below—add one point for each yes answer, and subtract one point for each no answer.

- Do you have gross sores on your groin, armpits, and neck that are oozing puss?
- Are they new?[232]
- Are you vomiting excessively?
- Do you have an intense, acute fever? Is it making you hallucinate? (Side question: Anything cool?)[233]
- Did you try to make friends with rats or rat fleas even after we told you not to?
- Are you vexed by a scorned witch who could be paying you back in kind for your cruelty?
- Have people commented that you look especially "blackish" or "deathy" of late?
- Would you say that you are exuding more puss than usual?
- Would you characterize yourself as "hated by God"?
- Did these symptoms kill you within two to seven days?

9 TOTAL POINTS: Sorry to hear about your death. Certainly is tragic. But while we have you here, just wanted to ask about life rights, since we're thinking about putting together this book . . .

---

232 Because if they're old, don't worry about them. But if they're new . . . yeah, you should worry.

233 "Tripping Bubonic" became a short-lived pastime among teenagers in 1355, and also 3003.

5–8 TOTAL POINTS: Bad news—you know that blood you're puking up? You really do need that. Skip down to "What to do if you've contracted the black death."

3–4 TOTAL POINTS: Congratulations! You've contracted some other kind of plague.

0–2 TOTAL POINTS: You should probably get those puss-oozing sores looked at,[234] but if you die in the next few days, it's far more likely that something else will have killed you.

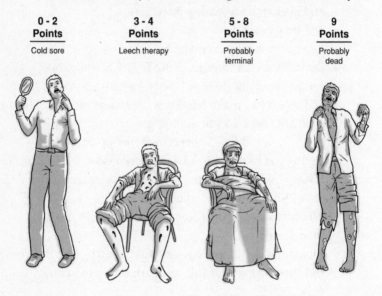

0 - 2 Points	3 - 4 Points	5 - 8 Points	9 Points
Cold sore	Leech therapy	Probably terminal	Probably dead

WHAT TO DO IF YOU HAVE CONTRACTED THE BLACK DEATH

- DO NOT jump into your time machine and burn for the nearest medically proficient time period. The future does not need your black death cooties.
- Clean up that puke/puss/blood.
- Send a message to a time-traveling friend via sub-space transmission, requesting some Black Death

234 Also, you should probably call the last few people you dated. It's the right thing to do.

B'Gone! circa 3003.[235][236]

- Apologize to that witch.
- Try to keep the sound of death throes down so as not to disturb your neighbors.

CRUSADING

The cornerstone of civilized life in the Middle Ages was crusading or being crusaded in the name of [your religion here]. It was common for practitioners of [your religion here] to be oppressed by the intolerant likes of [other religions here], regarded as inferior simply for holding different and equally implausible beliefs to be true. However, rallying together in their plight, [your religion here] took up arms and fought back—spreading the good word of [your God here] to all who would hear it, and especially to those who would not.

Chances are you will eventually encounter a crusade, join a crusade, be crusaded, or become engaged in crusade-related discourse that will result in your public and/or immediate execution if you don't tread lightly. Most territory, war, trade routes, sporting events, comedy, and occupations in the Middle Ages are either directly related to or exist under the influence of crusades and the religions driving them. To avoid being on the receiving end of a slaying in the name of [local crusader's God here], follow this quick dossier:

- As a rule of thumb, accept the religious views of those around you.

---

235 See footnote 233.

236 You do have a subspace transceiver capable of broadcasting irrespective of time coordinates, and a time-traveling friend, right?

Always Be Prepared to Accept the Most Popular Religion As Your Own

- Especially those around you with swords.
- After assimilating the most popular nearby religion, take it over the top. If you take [your newly chosen religion here] even more seriously than people who take it really seriously, they're going to have to take you seriously.
  - Self-flagellation and joining/starting a new religion based on your marital status are good ideas. As is screaming about [your god(s) here] in public.
- Be ruthlessly intolerant. Anyone who believes something other than you and the sword-wielding majority should be called out or converted.

- Murdering them as a show of faith may be called for if said faith is in question.
- There is no room for acceptance or lively debate in the Middle Ages.
- Travel to an unpopulated land and claim it in the name of [your religious empire here].
  - You may not even need to travel there.
  - And you probably won't even need to fight for it.
  - And you certainly won't have to convert the native population. In this era, news is slow to travel.

DRAGONS

Huge serpents that breathe fire, eat heroes, and generally cause havoc exist as myth in cultures all over Earth, as well as on at least one of Saturn's moons. Reports are sketchy, so we're erring on the side of caution.

WHAT TO DO IF YOU SEE A DRAGON

SEE SECTION "PREHISTORY: RUNNING THE HELL AWAY."

IF THE DRAGON IS RAVAGING THE COUNTRYSIDE

Did you not see the bit about running away?

IF THE DRAGON IS NICE AND TALKS LIKE SEAN CONNERY

In the event of a Sean Connery dragon, feel free to talk to it, befriend it, and ride it. Treat the dragon with the courtesy and respect you would afford Indiana Jones's dad, your elder, and a man/creature that could easily devour you and look/sound like the world's greatest badass while doing so.

IF YOU HAVE TO FIGHT THE DRAGON

Seeing as you're a time traveler, you might have certain abilities and responsibilities. It's not unlikely that you'll be asked to do battle with a dragon ravaging the countryside, especially if monster-slaying adventurer is your chosen occupation.[237] The following steps should help:

1. Wear really good armor.[238] Just ask somebody to borrow some. It's like asking someone to borrow a sweater in any other era.
2. You'll need something to circumvent fire as well as the dragon's kung-fu biting action. You might want to borrow one of those shields from the nearby knight guys. Get one that won't (a) burst into flames or (b) melt in your hands.
3. Remember that shotgun? This is why you brought it.

DID IT WORK?

It didn't work.

Sorry about that. Time for Plan B.

PLAN B—FINDING ANOTHER WAY TO KILL A DRAGON

1. Rack your brain attempting to figure out what technology would be useful for killing a dragon.
2. Leave the dragon's cave, castle, lair, etc., realizing that you can't kill it on your own.
3. Quest for a magic sword capable of defeating the dragon. If people you speak with tell you about a

---

237 See section "SURVIVAL GUIDE: EMPIRES: BLENDING IN."

238 See section "SURVIVAL GUIDE: MIDDLE AGES: FIGHTING."

Phil Hornshaw & Nick Hurwitch

sword with a goofy name, like Glumstig or Clownwedgie, you're on the right track.

4. Complete whatever trials are necessary to claim said sword.[239]

5. Quest to the far reaches of the realm to gather together a group of hearty heroes who will help you defeat the dragon. This fellowship should include:

   - An elf whose skills with a bow are matched only by how boring he is as a character.

   - A dwarf who is small enough to slip beneath the dragon but a little too small to be an appetizing snack. Judge candidates on length of beard, surliness of mannerisms, and general drunkenness (the more of all these, the better).

How to Fight a Dragon

239 If you're asked a series of riddles, remember that you'll need to know your favorite color beforehand. If your task is to remove a sword from a big hunk of rock, try the dynamite you brought with you.

- At least one "chosen one."
- At least one human warrior who's good in a fight but will probably betray you.
- At least one dice-rolling a-hole of a wizard who can do things like light up darkness and say profound-sounding shit to boost morale.

6. Once you've gathered your heroes, lead them to the dragon cave.
7. Stand behind the heroes. While they scream in agony as the dragon tears them apart, run beneath the dragon.
8. Stab the dragon in its soft underbelly, just where the neck meets the chest, using your quest sword, Phallusflayer.
9. Run away from the dragon cave before the dragon falls down or explodes or whatever dead dragons do, or before the remaining heroes get wise and decide to slap you around for using them as highly effective anti-dragon shields.

HOLY GRAIL

If you're ever bored or want to really impress a king or queen (perhaps as penance to avoid public execution), you can join the misguided thousands who have tried and failed to find the Holy Grail. A common preoccupation among bored adventurers and those tired of the rigors of crusading, the Grail is the cup Jesus used the last time he got drunk and asked friends to drink his blood.

However, wizard historians theorize that Merlin and his bastard dice-rolling wizard friends actually made up the Grail in order to derive amusement from non-magic simpletons. The wizards took bets on which seldom-bathed soldiers they could keep on the wild goose chase the longest.

So if you search for the Grail, don't plan on finding it.

And if you don't find the Grail, don't worry; you're in the same class as King Arthur.

But if you do find the Grail, you should put it back, as its presence in Medieval Europe could considerably alter the course of history.

First, though, shove it in Merlin's bastard beardface for doubting you, who, despite your lack of hygiene, found the thing he made up.

MEDICINE

They don't have Band-Aids in the Middle Ages. They have leeches. This is your first and only warning.

FIGHTING

If you've been trapped in the Age of Empires, or recall your careful study of the related chapter, you know the importance of "Pretend Swordsmanship." If not, review the section now and brush up on your skills in an area free of people to accidentally behead and of the cameras they may use to intentionally capture evidence of your subpar coordination, to be broadcast the universe over, an embarrassment that transcends time itself.

To recap: If you don't know how to win a sword fight, it is important to at least appear as though you can. Intimidation is often the first and last step in winning (or avoiding) mortal challenges. This, however, is a delicate balance. Sometimes your intimidation will merely provoke the man or warrior princess into wanting to fight you even more in order to protect his or her honor.

However, since the Age of Empires, a few things have changed:

HUGE ARMOR

Most warriors (knights) in the Middle Ages battle in standard leather plating, as well as chain mail and thick metal armor coveralls and helmets that protect nearly every inch of their bodies. This can help with not getting stabbed, as well as with the intimidation factor: You'll look about twice your actual size. However, being that the armor is metal and covers your entire body, you'll also weigh twice your actual weight—without adding to your underwhelming allotment of muscle. Even if you're capable of walking, you'll be slow and lumbering, and smell like the insides of a rotting pig carcass.[240]

For your purposes, armor better serves as a hiding place if you're ever sneaking through the inside of a castle.[241] Why, then, you might be asking yourself, would I ever want to battle in something so unwieldy? And you wouldn't have to, if it weren't for . . .

HUGE SWORDS

The heavy but somewhat practical swords of Rome have given way to heavier and impractical swords of the Middle Ages, which feel not unlike lifting an uncooperative bear. If you don't have armor on, these swords will slice you in half easily, but if you don't have a heavy sword, you'll never pierce the thick armor of another. Try doing three sets of fifty push-ups, five sets of fifteen chin-ups, and four hundred crunches every day for nine months.

---

240 Not just a rotting carcass, but the insides of a rotting carcass—the part that already smelled awful before the rotting began. That's how bad it smells in there. For a reasonable rate, however, your local blacksmith will happily cauterize your nostrils.

241 Note: Medieval armor takes roughly thirty minutes to put on.

Phil Hornshaw & Nick Hurwitch

THE FRENCH

As the Francs spread during the Middle Ages, so too does their laissez-faire attitude. They even came up with the term "laissez-faire," so the attitude they're spreading didn't even exist before they began spreading it. While also accustomed to heavy armor and heavy swords and using both to interior decorate, this developing power often shunned the notion of body protection and lifting heavy things, instead opting for thin, pointy swords designed for stabbing[242] and color-coded shoulder draperies designed to let you and others know whom to stab. This means four things:

- More mobility
- More stabbing
- More being stabbed
- Fast Frenchmen

If you find yourself unable or unwilling to operate heavy armor, being French might be right for you. Consult a medieval "doctor" before attempting to be French for three or more weeks, as results may be permanent. Rehydrate after leeching.

TIME MACHINE REPAIR

You have arrived, intrepid traveler: to a bold new era of human civilization. An era . . . with reliable metallurgy.[243] Blacksmiths, though still lacking in linguistic skill and basic hygiene, have developed advanced new techniques in their blacksmithing

---

242 As opposed to bashing, chopping, beheading, axing, or concussing.

243 Well, there was metallurgy in the Age of Empires, too, technically. But now the swords are bigger.

repertoire: better swords, formfitting metal armor, as well as other complexities designed to kill [enemy army here].

This doesn't necessarily mean that all or even most of the resources needed for the construction or repair of your time machine are available, but the laborious days of poop-gathering and Rube Goldbergering are over.

And with the presence of capable blacksmiths and an increasing interest in making things out of metal, your options have opened. The Middle Ages represent the first time in history that there is more than one way to repair your time machine:

HIRE A BLACKSMITH

1. If the damage to your time machine is merely structural or aesthetic, hire a blacksmith. Remember that disgruntlement is directly proportional to skill.
2. He's likely not keen to take instruction, so attempt to carefully make "suggestions," or delicately point to the desired result on your time machine schematics.
3. When interacting with your blacksmith, always keep an exit strategy in mind.[244]
4. You should be able to forge something "close enough."[245]
5. Work together with the blacksmith to install the newly forged components. If he seems suspicious, explain it away as a newfangled torture device.
6. Having befriended the dimwitted blacksmith through your increasingly amicable working relationship, offer a handshake before your departure, only to be unexpectedly wrapped in a stinky man-

---

244 Many a client has lost an eye or gained a cauterized hole as a result of a quick stab from a blacksmith's hot poker.

245 A surprisingly prevalent and acceptable result in time science.

Phil Hornshaw & Nick Hurwitch

bear hug. Blacksmiths have soft, squishy interiors beneath all that dimwittedness and filth.

7. Did you forget the schematics? Go back for the schematics.

BECOME A BLACKSMITH

1. If you're more of a loner, are predisposed to poor hygiene, or are incapable of controlling the irrationally violent outbursts of blacksmiths, take up as an apprentice.
2. Master blacksmithery.
3. Take on clients to maintain the guise of a legitimate business.
4. Stop bathing or caring for your physical health and appearance to maintain the guise of a legitimate blacksmith.
5. Consult your schematics and make the repairs yourself.
6. Fail.
7. If still alive, try again.
8. Return to your own time with a new, outdated skill set.

USE A TREBUCHET

1. If your time machine is structurally sound and merely needs to reach 88 mph (difficult to achieve due to a lack of fuel or paved roads), steal a trebuchet.
2. A trebuchet, like most technological advancements of the era, is a device of war designed to knock down the walls of, or kill, others. Similar in effect

to a catapult, a trebuchet can fire a live cow over three hundred yards in 8.3 seconds.[246]

3. Crank your trebuchet into firing position.
4. Put your time machine in the basket.
5. Get in.
6. Instruct the fifty or so soldiers and inbred convicts who helped you escape with the trebuchet to push it into position. Avoid aiming at nearby cliffs and castle walls, just in case it doesn't work.
7. Have your soldiers and inbreds bang their spears against the ground and their swords against their shields for dramatic effect.
8. While the sense of drama builds, crank the package into firing position using the cranks at the base of the device.
9. Yell "FIRE!"
10. If no one listened to you, cut the rope yourself.
11. Make sure you cut the right rope.
12. Activate your time machine as you soar through the air. Hopefully the physics you hammered out beforehand are accurate and your time machine reaches 88 mph in midair.

TAKE YOUR CHANCES WITH A BASTARD, DICE-ROLLING WIZARD

If you explain your situation to a wizard, he can whip you up a potion that will allow you to sleep until you're back in your time again. But don't sleep too long, as there's nothing less relaxing than waking up in an ancient castle in the midst of the apocalypse.

---

246 Congratulations, you've just invented ground beef.

VIKINGS (BARBARIANS/VISIGOTHS)

Following the fall of the Roman Empire and the veritable end of the Age of Empires, Europe was beset by these burly, seafaring ass-kickers. Vikings love to cruise the coasts of Western Europe, land at coastal villages, and leave them to burn in the destructive wake of their insatiable appetite to party. They are like rock stars of the late twentieth century but with more boats, less music, more swords, less drugs, and a similar amount of drinking. If you take to the seas during this time, you run the risk of encountering these early oversized pirates and villager-squashers. While it's best to avoid them entirely, with your luck you'll have to fight off several drunken boatfuls before returning home.

WHAT YOU NEED TO KNOW ABOUT VIKINGS

- Burly men of the sea, the Vikings are a force to be reckoned with on the ocean as well as on land. They're especially fond of both pillaging and sharp things. Avoid thatched-roof-cottage villages.
- You'll likely find Vikings attacking vulnerable settlements without warning and in force. When that happens, a strategic retreat is always a great option, though never by sea.
- Meeting a Viking in a pub or great feast hall is always better than meeting one right after they've come running off a boat toward the village you're in, ready to burn it down.
- Vikings find great value in drinking and feats of strength. You, being the literate, nerdy, machine-building, math-understanding time traveler that you are, will likely have little in common with them.
- However, if you have the capability to pound back

the mead, do so in earnest, and make lewd jokes. Vikings love a good "priest and rabbi walk into a bar" gag. Try to make friends with Vikings whenever possible.

- Vikings have simple motivations: partying, pillaging, conquest, and sexual conquest. If you have something to offer, you can often buy allies.[247]
- As early pirates, Vikings are as big a threat on the seas as in an unsuspecting village. They have little use for camels, however, so if you can just, you know, ride around on camels, it's kind of like Viking repellant. Just a thought.

WHEN FIGHTING VIKINGS

- Stick to what you know best: your shotgun. Use range, do your best to frighten your Viking enemies, but by no means should you do anything but buy yourself some time and then run.[248]
- In a hand-to-hand skirmish, a Viking will overpower you. Know any off-color jokes? Now's the time, as a Viking's sense of humor is one of his major weaknesses. They are also vulnerable to dynamite, if you happen to have any.
- Vikings are big fans of sharp things. Swords, axes, helmets, horns, tongues. If you are quick on your toes, stay low, run in figure eights, and dive through their legs whenever possible, they will

247 The authors of this guide suggest offering your purse rather than dropping your pants, but in a pinch, it's all better than a meeting the public relations end of a battle-ax.

248 See section "SURVIVAL GUIDE: PREHISTORY: RUNNING THE HELL AWAY."

Phil Hornshaw & Nick Hurwitch

eventually stab themselves or one another by accident or out of frustration.

- Vikings can sometimes be bought, although suddenly waving around a bunch of gold is as likely to get you robbed and killed as it is to ensure your safety. Always remember to mention that you have more money elsewhere and that cutting you in half is a deal breaker.
- Offer the Vikings musical instruments. The outlet of Heavy Metal is one that will quickly circumvent deep-rooted violence.

WIZARDING AND WITCHCRAFT

That's right—wizards. And not the Harry Potter and the hide-in-the-shadows-despite-being-all-powerful kind, either. We're talking the cut-the-heavens, turn-the-tides-of-battles, YOU-SHALL-NOT-PASS(!), Merlin-ass-kicking variety of wizards. The kind you should really avoid pissing off.

DEALING WITH WIZARDS

- First, assume that anyone who looks like a wizard is a wizard. Best-case scenario, they're just crazy. Worst case, your disrespect will earn you a place in Dumbledore's Army—of magical test subjects.
- If a wizard asks you for spare change, give it to him.
- If a wizard asks you to do something, you should probably just do it (at least until he can't see you anymore).
- Do your best to befriend/endear yourself to wizards rather than annoy them. A wizard friend is

great for dealing with people who would other-wise harass you. Also, dragons.[249]

Wizards May Be Medieval Hobos. But Better Safe Than Sorry.

IF YOU'RE FORCED TO FIGHT A WIZARD

- Don't let him talk, if possible—that's how they work their evil on you. Slap duct tape on him; stuff a sock in his mouth. Alternatively, you could hit him in the face with a pie.[250]

---

249 See section SURVIVAL GUIDE: MIDDLE AGES: DRAGONS, Finding Another Way to Kill a Dragon."

250 As clowns have not been invented yet, a wizard would not expect a pie to the face, whereas this is not an effective battle strategy in most other eras.

Phil Hornshaw & Nick Hurwitch

- Run up and steal that big pointy hat, which is the source of his power.[251]
- If you're fighting with a shotgun, try that.
- If you're fighting with a sword, hit that staff. That's the other source of his power.
- If none of that works, try running away[252] or begging for your life. Not all wizards make their enemies' brains explode in their skulls. Some just want a heartfelt apology.

WHEN YOU'RE ASSUMED TO BE A WIZARD (OR WITCH, OR WARLOCK, OR WHATEVER)

It seems inevitable that when time traveling in the Middle Ages, someone is going to see you working with spooky technology that has lights on it, or fire coming out of it, or parts that seem to move on their own. There are usually two possible outcomes to this scenario:

1. The person assumes you are a wizard and soldiers take you to the nearest king or warlord, where you will be forced to help him make weapons, or
2. You will be branded a witch or warlock, declared to be in league with the devil, and burned alive.

OPTION 1: Pretending you really are a wizard or witch: Remember that extra gunpowder we mentioned you should bring at the beginning of this chapter? This is where it becomes useful. Being able to create explosives for your warlord, even just to appease him while you pretend to make more, can be key in

---

251 This according to Walt Disney. Claim unsubstantiated.

252 See section "SURVIVAL GUIDE: PREHISTORY: RUNNING THE HELL AWAY."

avoiding being executed outright. Some other steps you should take:

- Always go out of your way to speak in riddles, answer questions with questions, and pretend you know more about what's going on than you actually do.
- Wear a pointy hat. Arrive precisely when you mean to.
- Make yourself indispensable to whomever you're with by offering to blow things up for them, or by threatening them with your "magical might."[253]
- Maintain an air of mystery and hermit-atude: long beard, flowing robes, white hair, scraggly appearance—all these things might make you look like you have some kind of brain damage in another period, but here they are essential for spooking others into not murdering you.
- Be sure to do some magic every once in a while. Flip on a flashlight, threaten someone with an electric razor, use a lighter. Make sure everyone thinks you could set them ablaze with a mere thought.
- Don't overdo it with the threatening—be sure to be an "ally" wizard or witch and not an evil "top of the tower must be killed for the good of the land" kind of wizard or witch.
- Convince others to help you with promises of riches and power, and use their manual labor to repair your time machine.

OPTION 2: Avoiding being burned alive as a wizard or witch:

---

253 "Magical might," in this case, refers to your shotgun again. Keep that thing handy at all times.

Phil Hornshaw & Nick Hurwitch

One major downfall of being branded a magic-user is that it often carries with it the title of "heretic," or some such religiously infused and equally damning name. Witches and wizards/warlocks are often believed to be in league with the devil, and often denying you're a witch or warlock will be taken as evidence that you are one. The tests for determining such—prolonged submergence in water and burning—aren't so favorable, either. The following is what to do if you're branded a witch:

- Cackle maniacally, firing your shotgun in the air, as you retreat to your time machine and scoot off to find another era to mess with. Threaten your inevitable, prophesied return.
- Vehemently deny that you're a witch or a warlock, while simultaneously explaining that you're a time traveler from the future and that you've come to make everyone's life better. Tell them you'll teach them to blow things up. Then blow something up, like the people you just explained time travel to, who are now witnesses that need to be dealt with.
- Accuse the person accusing you of being a witch of being a witch. That sometimes works.

IF YOU'RE ABOUT TO BE BURNED AT THE STAKE:
- Know that you'll be remembered. As a cautionary tale.

IF YOU'RE ABOUT TO BE SUBMERGED IN THE RIVER FOR WITCH-TESTING:
- Wait till you're in the water, then scream, "There's something in here!" and fake a shark attack. (You may be in a cage. That's fine. It's an invisible shark and you're drawing on high school drama here.)
- When the bewildered villagers stop lowering you

and begin to watch, horrified, abruptly end the attack.

- As they look on, questioning, tell them you were "just attacked by the devil." Point at your executioner (or king, or random soldier) and say, "He was looking for you!"
- As they interrogate their former cohort, swim to the other side of the lake you're in. Steal the unattended horse on the beach and ride off into infamy/obscurity. They often won't give chase: There are always other alleged heretics to burn.

SURVIVING IN TIME: INDUSTRY

1300 C.E.–1940 C.E.

(Advent of Guns to Advent of Bigger Guns)

HOW TO KNOW IF YOU'RE IN THE DAWN OF INDUSTRY
- Pollution
- Guns
- Black lung
- America
- Colonialism
- Rampant literacy, the desire for
- Transcontinental warring
- Penicillin
- Exploration

WHAT YOU SHOULD BRING
- This guide
- Boomstick
- Running shoes
- Sack of potatoes
- Gas mask
- Gunpowder/dynamite (or other explosives)

- Backup time machine battery
- Backup time machine

INTRODUCTION

This era of Earth's history is as expansive as it is diverse. Before the calming, ubiquitous nature of globalization but still during the erratic, violent age of colonialism, nearly every corner of the globe undergoes rapid change.[254]

However, this rapid change also makes the Age of Industry the most nebulous for a time traveler to prepare for. You could just as easily be desperately attempting to eat rice with chopsticks in Japan as you could be attempting to remove the rather snug corset off a dehydrated saloon wench in the Wild West; just as likely to be struggling not to inhale toxic condiment gas in a World War I trench as differentiating navy blue from dark gray uniforms in the fading dusk light of a horrifically gory American Civil War battle.

Your best odds of survival are to blend in with your surroundings, try not to stir any trouble, and, when in doubt, resort to your pump-action Boomstick. If it's not in here, you don't need to know it, and if you still think you need to know it, you should be somewhere else. Good luck.

RENAISSANCE (1300 c.e.–1600 c.e.)

Following the Middle Ages, at least in most of Europe, is the Renaissance—a time of intellectual growth and also heinous, anti-intellectual torture and execution. A fun time to visit, but

---

254 Though by now, everyone has more or less accepted that the Earth does not, in fact, have any corners. Don't be the one idiot still adamant that the world is flat.

you wouldn't want to live there, and you certainly don't want to die there.

If you're traveling in Italy and surrounding areas between the fourteenth and seventeenth centuries, you're likely to hit this movement. It involves the advent of guns, the continued use of swords and horses, some light astronomy, cannons, war, and sculptures of naked people. Plan accordingly.

ASTROLOGY

You have a time machine and therefore the ability to see the future. What are you going to do with fortune-telling? We'll tell you what: scam people. You no longer need to be imbued with "The Sight" or even be able to differentiate that crab thing from that goat thing. Just remember to keep the readings on the cold side lest you muck up history with your profound accuracy.

Things you need to know about astrology:

- It's not the same thing as astronomy.

ASTRONOMY

The beginnings of an understanding of celestial bodies, the solar system, and the universe are burgeoning during the Renaissance. Science is first taking hold among humanity, and it's just so damn cute. Unfortunately, this causes some pretty heavy backlash among the "God made it that way" crowd. And that leads to burning, torture, and hanging. Not as cute.

Things you need to know about astronomy:

- The less you know, the better.
- If you do know something about astronomy, and you're not interested in being branded a heretic and potentially burned, you're going to want to shut up about it.

- Important astrological realizations of the Renaissance that did not immediately take hold: The Earth is round and the Sun is the center of the universe. Feel out the popular opinion before guffawing in one direction or the other.

Historical Figure Spotlight

DA VINCI

Leonardo Da Vinci is a painter, a sculptor, an inventor, a writer, a scientist, a scholar, the namesake of a Ninja Turtle, and an all around show-off.[255] He is regarded as perhaps the most talented man who ever lived.

However, many time travelers have muttered under their jealous breath, "Like to see how much he'd pull off if he was born after the invention of the Internet." Though with all due respect to Al Gore, Da Vinci probably invented the Internet, too. He simply didn't live long enough to build a proper infrastructure or the means to superimpose text over photos of hilariously cute domesticated cats.

Contrary to popular belief, Da Vinci does not always look like an old guy. It's just easier to appreciate someone so much better than you at everything if they appear old and wise. Da Vinci in fact spends most of his life as a thin, antisocial weirdo, sleeping only in power naps in order to maximize the time he spends outdoing nearly everyone else in history.

If you really want to blow a historical figure's mind with the future, or are in need of an excellent prop for your final history exam, Da Vinci is easily the best choice.[256]

---

255 His involvement in some type of hidden "code" is purely speculative, and discussion of such should merely be taken as an indicator of one's literary tastes or intellectual worth.

256 Even if you will look like a sentient pile of dog turds in comparison to Da Vinci. Worth it.

Phil Hornshaw & Nick Hurwitch

BLENDING IN

Much of the class system present in the Middle Ages persists into the Renaissance. Here, though, there are also a good deal more philosophers, poets, musicians, painters, and other artists—in other words, if you've got ambition to be a slacker, start a band, or "just, ya know, work on my art," you'll fit right in.

SCIENTIST—An effective way to hide your time traveling past and time travel repair ambitions. Has the serious drawback of making you a target for Inquisitors.

INQUISITOR—A handy job if you consider yourself of the "evil time traveler" persuasion, and also gives you access to the confiscated scientific materials of the scientists you are currently stretching over hot coals.

MERCHANT/BANKER—The Renaissance saw the creation of an economic middle class of people who made money by selling stuff without actually having to make or farm it. It's a pretty sweet gig, and keeps funds fairly liquid for spending on replacement time machine metals and components. Banking is even cushier. Merchants and bankers generally aren't targeted by Inquisitors, which is also nice.

PRIEST—Possibly the most corrupt figures throughout the Renaissance, and that's including the guys who go around stabbing people for money. Many members of the Catholic Church collect a ludicrous amount of money from the population by engaging in such questionable practices as selling indulgences. This spawned events such as the Great Schism and the Reformation. As a priest, you can sell indulgences, too! And you can use the money to buy yourself some nice things to take back to your present with you. It really only costs you all moral dignity and possibly your immortal soul.

ENLIGHTENMENT (1600 c.e.–1850 c.e.)

For some reason the Age of Enlightenment includes kind of a lot of popular uprisings. Your best bet, in general, is to avoid period-specific clothing that suggests you are (1) a loyalist to King James of England, (2) a rebel against King James of England, (3) any kind of French aristocrat, (4) a colonizing soldier from another nation. These are the kinds of people that other kinds of people tend to round up and execute during this period.

If you're hanging around the Enlightenment, do some reading and catch a few plays, as there are some good brainy ones coming out. Other than that, a powdered white wig is basically always a winner for defusing any questions about your 2340 human standard haircut mohawk.

AMERICAN REVOLUTIONARY WAR (1775 c.e.–1783 c.e.)

Eventually, the thirteen British colonies of America decide they have had enough of being ruled by a figurehead from across an ocean, taxation without representation, and drinking tea with pinkies out. The result is the Declaration of Independence and the formation of the self-proclaimed "Greatest Nation Ever."

Blending In

- If you have a British accent, drop it (and how).
- Take advantage of the "stand still, shoot the guy three feet in front of you" brand of warfare by bringing a shotgun.
- Take advantage of the "stand still, shoot the guy three feet in front of you" brand of warfare by dodging. No one ever dodges.
- Red: The Worst Camouflage.
- If you're smart but kind of ugly, hole up with Ben

Franklin. He may look like a hairy foot, but he is loved like a new handbag by the ladies. Ask for some tips, or observe and duplicate.

- Swing by the capitol and give Mr. "Bear Arms" Hamilton our love.[257]

FRENCH REVOLUTION (1789–1799)

At this turning point in political history, the grumpiness of the French gets the better of them and they overthrow the rule of a centuries-old, tried-and-true monarchy in a few short years. The revolution paves the way for democracy, the long, slow battle for inalienable human rights,[258] blah blah blah: it looks nice in the history books, but the reality is that this period of French history gets a little hairy (and not just because Frenchwomen refuse to shave their legs).

Blending In

Here are some useful tips and phrases that will keep you in with the revolutionaries and out of the guillotine:

- Don't wear fancy clothing. The aristocracy and their overzealous fashions are the enemy!
- **"VIVE LA RESISTANCE!"**—a phrase that the revolutionaries may or may not have actually said, but will surely earn you points if it ends up catching on. Not to be said near people in fancy clothing.
- If someone offers you cake, eat it. Fairly certain that's how the story goes.
- **"VIVE LE ROI LOUIS!"**—useful if you're being carted off, rounded up, or forced into a long line at the

---

257 He prefers for our love to be given in the form of fist bump explosions.

258 Which hold up pretty well until the aliens show up and alienate them.

guillotine. Not useful for dying with any dignity, you traitorous swine.

- Make a funny face just as the giant blade flies downward at your ground beef–like neck—it'll stay that way once your severed head lands in the basket. Quite comedic.
- When in doubt, punch a dude at random or light something expensive on fire. This is how rioting is done, and the French practically invented it.

DEALING WITH PEOPLE WHO CAN KILL YOU

Pirates

As colonialism ramps up and European nations take to the seas, goods begin to be trafficked heavily over the open oceans. And whenever something valuable is being moved from place to place, inevitably, it will bring those who would steal it. Enter the scourges of the sea, those flamboyant scoundrels: pirates.

You may think, "Those guys with the parrots? Why would water-bound cross-dressing animal lovers be dangerous?" And it's precisely that attitude that will get you splayed by a rusty blade.

Should you find yourself in the era of colonialism with a desire to travel, you may find yourself on a ship. And every time you take to the seas, you could face piracy—especially if you're having your seventeenth-century spring break in the Caribbean.

What you need to know about pirates:

- They will mindlessly and endlessly search for and then hide treasure. If you throw something shiny, a pirate will chase it in all circumstances.
- Pirates are Earth's first cyborgs. Often you'll find a pirate has lost many a limb to cannon fire, angry sea life, and angry bar fights over how many and to

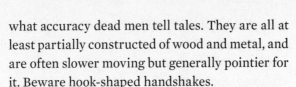

what accuracy dead men tell tales. They are all at least partially constructed of wood and metal, and are often slower moving but generally pointier for it. Beware hook-shaped handshakes.

- All pirates are almost constantly drunk. This makes them both incredibly less and incredibly more dangerous. Avoid snarky comments about their puffy shirts and illustrious parrots.

- A life of endless horizons and unforgiving sea has a tendency to increase depression. A pirate often only needs a friend to listen and a non-parrot-occupied shoulder to cry on.

WHEN FIGHTING PIRATES

- Remember that peg legs, eye-catching hats, long, heavy swords, limited depth perception, rampant drunkenness, and intense scurvy make it hard for pirates to engage in land pursuits. Run.[259]

- Every pirate has scurvy, which is caused by an acute deficiency of vitamin C. Bring a glass of orange juice and turn an angry enemy into a new friend. A friend with a boat.

- If you're unable to quickly shoot or stab a pirate, try to spook the parrot. All pirates' shoulders are home to parrots, and aiming and stabbing are exceptionally difficult with an angry wing-flapping monster biting and clawing at your shoulder.

- Most pirates wear eye patches not because they look freaking awesome,[260] but for the express purpose of navigating and fighting people in the dark-

---

259 See section "SURVIVAL GUIDE: PREHISTORY: RUNNING THE HELL AWAY."

260 Although eye patches are prevalent as a fashion statement of badassery in any era.

ened holds of ships. Expect a pirate to have strong night vision. Distract the pirate or defuse the situation by offering rum.[261]

- Grated floors, crocodiles, and planks of wood are natural enemies of the pirate. Make use of these, both for physical aid and intimidation, when engaging a pirate. Wave a two-by-four when desperate.

WHEN CAPTURED BY PIRATES

- Pirates' preferred means to dispose of prisoners is known as "the plank": a hunk of wood that hangs over the side of the ship, which bound captives walk until they fall into the swelling ocean below.[262] There are a few things you can do to help you survive such a fate:
  - Try your damnedest not to go into the water with a rope binding your wrists. Ropes tend to swell in water and you'll be impeded to the point of drowning.
  - Whenever possible, employ the help of fairies, which are natural pirate enemies.
  - Claim to your captors that you're worth a king's ransom, then compliment the lead pirate's lustrous and luxurious beard.

KEY PIRATE PHRASES:

- **YAARGH**—"Oh really? You don't say"; "That's water"; "C'est la vie de les pirates"; "Can you direct me to the nearest library?"

---

261 Pirates love rum. Also the phrase "Yo ho ho."

262 "Walking the Plank" can also refer to a sex act common to pirates and some late twenty-second-cenutry uraniumcore pornography, which we are legally prevented from detailing here. Trust us when we say that you'd rather have the one where you fall into the ocean.

Phil Hornshaw & Nick Hurwitch

- **ME HEARTIES**—"I love you guys."
- **DAVY JONES'S LOCKER**—a poorly kept bathroom.
- **KING'S RANSOM**—$26 U.S. (2020 adjusted).

Ninjas

Allow us to settle an age-old debate: After intensive scientific study that took about two hours, it has been concluded without a doubt that ninjas are better than pirates. The only way a pirate can kill a ninja is if the pirate happens to find the ninja unconscious, or if the pirate sneaks up behind the ninja while the ninja is fighting off ten other pirates simultaneously. And in both the above scenarios, the pirate is far from a guaranteed victor.

What you need to know about ninjas:

- Ninjas are everywhere at all times. Anyone you pass on the street could be a ninja. Empty streets, potted plants, and light fixtures are almost surely ninjas. This is a function of their role as secret assassins, revenge seekers, and security guards. Be paranoid, because ninjas are probably tracking you right now.
- Unlike samurai, ninjas are totally without honor and care nothing for the lives of the people they kill. They'll kill you for using chopsticks wrong or because you irritate them with your future sensibilities. But you'll never know it was them.
- Usually ninjas are either born into ninjadom, are orphan street urchins that are exploited by ninja masters and taken under the protective wing of ninjadom, or willingly enter ninjadom in order to seek revenge. If any of these apply to you, you, too, can become a ninja, which may be your only real chance at defeating a ninja.

- The hell are you doing fighting a ninja?
- You've probably already lost.
- If you haven't yet lost, do NOT allow a ninja to steal your time machine or discover your time traveling secrets. The only thing worse than an evil time traveling copy of yourself is an evil time traveling ninja.[263]
- Don't try and shoot ninjas, as their swords can deflect bullets. You'll probably end up shooting yourself.
- Don't try to swordfight them,[264] as their speed with a blade is unrivaled.
- Ninjas can instantly detect weak spots and exploit them for easy victory. If your entire body is a weak spot, you may actually delay your death by precious seconds.
- Cyborg ninjas are the worst kind of ninjas. Robocop backup recommended.
- Ninjas are incapable of love. If you manage to kiss one, you may overload its logic circuits. You may also get your lips cut off.
- For ninjas, secrecy is everything. Therefore, if you can recognize a ninja's non-ninja identity, you can easily defeat him by screaming, "This ninja is Hiroshi Fitzsimmons!" He might cry and run home. More likely, though, he will murder you and everyone else in earshot.
- Have a ninja murder your closest friend, your parents, or your one true love. Fueled by revenge, become a ninja yourself and seek out the bastard ninja who murdered your friend, parents, or true love. The only thing stronger than a ninja is a revenge-fueled ninja.

---

263 See section in Chapter 5, "Timebattling Your Time Self."

264 See section "SURVIVAL GUIDE: EMPIRES: WAR, Pretend Swordsmanship."

- Ninjas take no prisoners.

Samurai

Think knights, but in Japan, and in skirts. Okay, they're not skirts, they're traditional kimonos or something. Samurai are dangerous with a sword, but won't even know about guns until the late nineteenth century—when they're shot to death by those guns, pretty much uniformly. Samurai are generally soldiers loyal to a specific warlord, whom they fight for and who pays them, or is the elder in their village. They probably won't pick a fight with you, as they are a little less douchey than the soldiers of most other cultures. However, if you do run up against one, the following will help you to get through the encounter.

WHAT YOU NEED TO KNOW ABOUT SAMURAI:

- They dress weird, have weird hair, and carry two swords. Jokes about the former two result in introduction to the latter two.
- Samurai have a strict code of honor that demands the blood (or heads) of enemies for a variety of seemingly not-that-big-of-a-deal reasons.
- The strict code of honor also demands that a samurai disembowel himself should he dishonor himself in any of a variety of ways. Seriously, they'll cut their guts out for, like, no reason. Try not to get any on you.
- You can sometimes hire samurai to work for you, provided you have feudal Japanese money or something to trade, like a mystic staff that opens portals through time, or demon warriors that look not unlike giant turtles. Ninja turtles.

- Many samurai are actually pretty cool, and they'll fight for you provided that you, too, are pretty cool.
- Samurai should be especially feared in groups of seven or greater.

They're Not Skirts, But They Might As Well Be

WHEN FIGHTING SAMURAI

- You brought your shotgun, right? Because these guys really do not have any concept of "exploding swords."
- If you cut a samurai's hair, he will commit suicide immediately, or become **ronin**, a dishonored, masterless samurai.[265]
- If you're a ninja, don't even worry about it.
- Samurai fear strong upward gusts of air because they don't wear pants. Use your environment to pray on the pantless weaknesses of your enemy.

WHEN CAPTURED BY SAMURAI

- Gain their respect by having great honor.
- Disarm their better judgment by displaying kindness and rudimentary magic tricks to their women and children.
- Defeat the angry one in a spar. Then, you will be one of them. Hope you like skirts and hair ties.

INDUSTRIAL REVOLUTION (1850 C.E.–1940 C.E.)

As factories and automation pick up, the world enters the period for which the Age of Industry earned its name. Things are a mite more civilized: The age of pirates comes to an end, and the age of child labor, strip mining, Old West gunfights, and corporate greed comes into vogue.

On the plus side, in much of the Western world at this point, it gets a bit tougher to get murdered on general principle (depending on your skin color and geographic locus) or for your clothing or apparent political leanings. There are other immi-

---

265 All bets are off when it comes to those *ronin* guys. They're not bound by the honor rules of normal samurai, and you can't defeat them just by cutting off their hair, even though it'll totally ruin their day. Use caution.

nent dangers, however—like getting roped into pre-OSHA menial labor.

BLACK LUNG

An ailment commonly incurred by chimney sweeps, factory workers, and peasant children forced to jump-start their careers at age six, black lung disease is a filling or coating of the lung's alveoli with dirt, soot, fecal matter, and other toxic materials black and powdery in nature. To avoid black lung, don't go to London in the nineteenth century. Alternatively, avoid chimney sweeping, factory labor, and being an orphan.

BLENDING IN

Here are a few tips for casting off suspicion during the Industrial Revolution:

- Get dirty. Everyone in this period is covered in coal dust 109 percent of the time. Just roll around in it. There's no better way to avoid pesky questions like "How come you still have all your fingers?" when it seems like you just haven't gotten around to losing them yet.
- Practice being put-upon by the wealthy elite.
- Avoid prostitution, prostitutes, or creepy men seeking prostitutes, specifically in White Chapel, London, England, circa 1888.

DEALING WITH PEOPLE WHO CAN KILL YOU

Most every participant in the Industrial Revolution is too worried, for the first time in history, about "makin' a quick buck" and "climbing the corporate ladder" to concern him- or herself with murdering you outright. The real killers here are of the slow-burn variety: economic inequality, manual labor, and inhaling too much soot.[266] So, don't work at a factory, definitely don't work as a chimney sweep,[267] and otherwise keep your boomstick handy and your head on a swivel.

CIVIL WAR (1861 c.e.–1865 c.e.)

Around this time, the young nation of America realizes that it doesn't quite agree on that Constitution thing everyone has been going on about and decides to fight itself.[268]

If you happen to find yourself in the midst of the Civil War, pick a side.[269] Then, go AWOL. This really isn't a war you want to fight. If you're pro-Union, you'll find that it's really a lot less about slavery than you realized; if you're pro-Confederacy, you'll find that trying to keep slaves who want you dead isn't all it's cracked up to be.

The warfare is also bloody and ruthless, and comes with a high body count. Ambushes in forests, battles on hills, and stabby implements on rifles are all commonplace. Most importantly, you can't change the outcome of the Civil War. When the South finally does rise again in 2033 (dubbed "Heehawzapaloozaa"), things don't go so well.[270] Plus, no matter how right you

---

266 See section "INDUSTRY: INDUSTRIAL REVOLUTION: BLACK LUNG."

267 Did you know that if you get stuck in a chimney, they might not find your body for *years*? Surely a time traveler can figure out a more dignified way to kick the bucket.

268 This delighted the British, who were still pretty sore over getting their colonial asses handed to them a hundred years prior and figured they would have used the land for something more dignified, like polo, or fox hunting.

269 Your side may be chosen for you based on your ethnicity, the color of your shirt and how you feel about cotton farming.

270 The resistance is quashed in less than three hours after accidentally marching through the birthday party of an on-leave National Guardsman.

think you are, a civil war feels not unlike a hostile divorce. Except, in most instances, for all the bloodshed.

BLENDING IN

In general, hiding your political geographic affiliations is key to blending in throughout the American Civil War, as where you are from (or seem to be from) has a lot to do with whether people decide to shoot you. On the whole, you're better off in Canada.

Practice your Southern/Northern accent. Or better yet, pretend to be French. Americans are still friends with the French at this point, and prior to the Great Wars, being French is still considered "hip" and "not cowardly." Gray and blue clothing should generally be avoided. It also helps to avoid seeming like you have any food to spare should you run across any soldiers, as being a soldier sucks and those guys are hungry. If you're fat, attempt to appear less fat.

Conversely, having lots of food can make you some friends among soldiers, which is good until soldiers from the other side show up.

Don't be afraid to use whatever set of soldiers you've become friends with as human shields. If that doesn't work (see below in "Dealing with People Who Can Kill You") your next best chance of blending in is to avoid blending in altogether. Take the semi-public transit system known as the "Underground Railroad." We're not sure how it works, but it would appear that it's pretty self-explanatory.

DEALING WITH PEOPLE WHO CAN KILL YOU

Dumbass Soldiers

Regardless of what you might think of the Civil War conflict, the reality is that both sides consist of a lot of people with little to no soldier training, many of whom don't even have uniforms. The Union troops are generally a little more together than the Confederate troops, but you're not going to want to piss off anybody whose big political sticking points are "forcing other people to be my friend" or "forcing other people to do the hard work for me."

Avoid:
- **THE STABBY AND SHOOTY ENDS OF THE GUN**— Those are the same end.
- **CANNONBALLS**—They would likely hurt less if they just went ahead and exploded. Instead, they remain solid balls of limb-removing metal.
- **PEOPLE WITH BIG MUSTACHES**—Can't be trusted.

WILD WEST (1865–1910)

Following the Civil War, the United States was a bit of a mess. Much of the country had been ravaged, and that was especially true of the South, where one Union general just went around burning shit. So a lot of people, ex-soldiers and others who had nothing better to do, headed west.

When it comes to the West in the latter half of the nineteenth century, you should note that the moniker "wild" is there for a reason. It's also important to understand what meaning of "wild" is implied in the usage. This is "wild" as used in the following sentence: "That wild animal is tearing Billy's face off for no good reason." It is NOT "wild" as used in this sentence: "That wild coed is tearing her shirt off right in Billy's face for no good reason."[271] Please make sure you understand this key difference before visiting the Wild West.

---

271 Though if that does happen, there is a whole niche market of porn in the twenty-first century dedicated to "Past Boobs."

BLENDING IN

Looking like you belong in the Wild West isn't too tough. Get dirty, wear pointy boots, put on a hat. Pink is to be discouraged. Browns are in this year, and every year, in the Wild West. If you're a woman or have always wanted to be a woman, get fitted for a corset by pulling the strings as tightly as possible. Tighter. Slap a ruffled dress over top, despite all the places where dust can get caught, and you're a regular damsel.

It can also help to look like you know what you're doing with Wild West–appropriate objects and skills: Try twirling your pistol on your finger and maintaining the outward appearance that you're not completely terrified of your horse.

SOME OTHER THINGS YOU CAN DO TO LOOK LIKE YOU BELONG IN THE OLD WEST:

- Farm
- Ranch
- Drink whiskey
- Play poker (but not too well) (actually, play badly, just to be safe)
- Obsess over gold
- Learn to remove a corset with one hand
- Drink more whiskey

DEALING WITH PEOPLE WHO CAN KILL YOU

Cowboys

Any adventuring done in the West of the United States after the Civil War to roughly the advent of the automobile will likely run you up against cowboys, outlaws, ruffians, and scoundrels of the six-gun-toting variety. For some reason, this era has a great deal of appeal to time travelers of many different stripes, and their lust to experience the "Wild West" often results in them getting "fucking shot in the gut," after which they bleed out slowly and painfully. If you're headed to the Old West, be prepared to deal with some bad-news dudes.

WHAT YOU NEED TO KNOW ABOUT COWBOYS:

- All cowboys are adept horsemen but don't run so well due to saddle sores. They make up for this by bringing ~~victims~~ adversaries to them using their cattle-roping skills.
- All cowboys have an innate psychological fear of machinery. It relates to feelings of inadequacy.
- Cowboys hate to lose, and they have a strict morality when it comes to card games and cheating. They enforce this morality by shooting cheaters. They have also been known to kick over tables, gather up their cards, and go home in tears.
- All cowboys have haunted pasts. This makes them drink heavily. You can make a cowboy friend and ally with whiskey—he might not be all that useful, but at least he won't be shouting, "Dance, runt!" as he fires bullets into the floor at your feet.[272]
- A cowboy cannot resist gold, especially if (a) he cannot see it, (b) he will probably have to dig in the earth to obtain it, and (c) its existence is questionable. This also applies to oil.
- Fool's gold is an easy way to make a cowboy instantly friendly, and also an easy way to get shot by a greedy but stupid cowboy.

WHEN FIGHTING COWBOYS:

- A cowboy cannot resist shooting at objects thrown into the air. Waste his ammo by emptying your pockets.

---

272 When in doubt, moonwalk. It confuses and frightens cowboys.

- Cowboys' pistols only fire six shots. Keep track. You wouldn't want to have to ask yourself, "Do I feel lucky?"
- Distract a cowboy with fool's gold, or just offer it to him. This will generally placate any cowboy, provided he doesn't think you have more.
- When confronted by a gunslinging cowboy, the best solution to any disagreement is to just shoot him first, preferably while he's walking into the middle of the street to have his duel with you, or walking away to get into position for said duel. Being a live, cowardly time traveler outpaces being a dead, honorable cowboy.
- Steal a cowboy's hat and he will be rendered powerless.
- Catch a cowboy's lasso and his head will explode.
- Insult a cowboy's horse, and he will react by becoming depressed, often causing him to withdraw from any battle for lack of confidence.
- Baked beans render cowboys inert and docile.

WHEN YOU'RE CAPTURED BY COWBOYS:

- Do your best to convince them not to hogtie you.
- If they hogtie you, do your best to convince them not to leave you on some train tracks.
- If they leave you on some tracks, wait till they leave, and then roll off them.

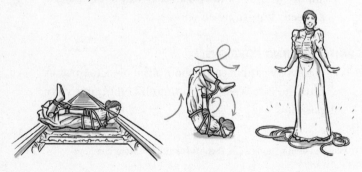

Phil Hornshaw & Nick Hurwitch

Things That Suck About the Wild West:

DESERT—That's mostly what it is. Really hot during the day, really cold at night, an intense lack of rain, and a good place to get lost and die. Did you know when you die of dehydration or heat exhaustion, you go crazy right at the end? You eat sand, gnaw on your own limbs, try to beat rocks into submission for looking at you wrong, and generally slip into unconsciousness, delirious with pain, heat, and sand in your ass.

COWBOYS—See section above "Dealing with People Who Can Kill You."

GOLD—Many settlers are compelled westward with their families and brave dysentery, dead oxen, and broken wagon axels in hopes of discovering gold and living the rest of their lives on Easy Street. But, of course, in the middle of nowhere, there are no streets. Most of these poorly prepared saps don't want to do any actual mining, so they just sift the rocks at the bottoms of streams and eventually starve their families to death. There is no free lunch hidden at the bottoms of streams.[273]

Because everyone is so gold-crazy, however, if you happen to have some, it makes a damn good bartering chip. Be careful not to get robbed, though. Just about anyone out here will rob you, including the horses and some saloon doors.

(THEY'RE NOT ACTUALLY) INDIANS—The new enemies as White Devil settlers push westward are the Native American tribes that apparently already live there. The whole thing ends up being rather tragic, but for practical reasons,

---

273 Fact: There are no leprechauns in the Wild West.

you should know that if you look like the settlers they're struggling against, these guys may cut your skin off and then kill you. The skin-cutting part happens while you're still alive. Steer clear.

PRETENDING TO KNOW HOW TO GUNFIGHT

In the Wild West, the only things less common than baths are kind dispositions. Chances are, if you're kicking around the West, you're going to piss somebody off. And that person will likely threaten you with some sort of firearm, or a drinking contest and then a firearm. You'll need to be able to pretend like you know how to fight back.

1. Bring a gun—This is key.
2. Point it at the guy who you want to shoot—The idea behind intimidation with a firearm is to actually show you know which end is the scary one. Alternatively: Aim it at the chandelier just above the guy you want to shoot.
3. Pull the trigger—Try not to freak out and drop the gun when it makes a loud noise.
4. Run—This is especially important if you missed or the other guy didn't immediately die of his chandelier-related injuries. Remember—a zigzagging target is harder to hit.

WORLD WAR I (1914 c.e.–1918 c.e.) AND EARLY TWENTIETH CENTURY

The worst of all early twentieth-century wars, World War I saw some pretty grievous atrocities: trench warfare, naval battles, gassings, and the advent and subsequent horrific use of the machine gun. Warfare often boiled down to some jackass ordering a bunch of soldiers to just charge at a machine gun and get shot by it until someone was lucky enough to human shield his way through and shoot the gunner.

When adventuring through Europe in the mid-1910s, you're going to want to avoid most of France and Germany and specifically anyone wearing a helmet with what looks like a spear coming out of it, as those guys aren't so nice. If you get roped into storming an enemy position, just be sure you're the last one off the boat/jeep/Trojan horse.

MUSTARD GAS

Used to great (and horrific) effect in the Great War, mustard gas is one of the first examples of chemical warfare. Not to be confused with the delicious twenty-first century gastronomic condiment commonly inhaled with pastrami milkshakes, mustard gas of the Age of Industry is not something you want inhale or put anywhere near your sandwich.

GUNS

You'll begin to notice that that shotgun and related arsenal you've been toting around with you is now less of a convenient upper hand and more of an absolute necessity. Whether you're caught in war, are goofing off in the Wild West, or are decimating entire cultures still unmolested by searing, high-speed metal, you're going to want to have a gun or keep close company with those who do.

BIGGER GUNS

Even if you have a gun, the playing field begins to diversify throughout this era. Now, in addition to the standard short-range war rifle, there are smaller guns designed to be portable and fired with one hand (pistols), huge guns designed to hit several targets at once or several in a short period of time (Gatling, machine), and guns designed to shoot large incendiary rounds or solid balls of iron (cannons).

CARS

Good news! Around the turn of the twentieth century, cars are invented. Bad news! They're pretty awful. Why you should avoid early twentieth century cars:

- You can probably run faster than most of them.
- Horses can definitely run faster than most of them.
- Most have to be hand-cranked before use: not ideal for getaway situations.
- You'll need some kind of anti-bug-in-mouth device, as windshields at this point are still considered a foolish commodity.[274]
- You can't use them to rob, hijack, board, or race trains.
- Cobblestone streets: They'll give you a permanent stutter and a bonus rib contusion.

Cars make big strides after World War II, when things like air-conditioning, shock absorbers, and engines more powerful than horse legs are in vogue. Until then, you're better off putting up with saddle ass.

TIME MACHINE REPAIR

With technology chugging along like so many Wild West locomotives, the ways to repair your time machine are now nearly as plentiful as the ways to break it. Coal! Steam! Die-casting!

---

274 Does your horse have a windshield?

Phil Hornshaw & Nick Hurwitch

Slaves![275] With the notable exception of duct tape,[276] nearly everything you need to patch up your oft-broken time traverser can be found in this era.

But because we can already hear your nasally demands for clarity, here are the QUAN+UM-approved repair suggestions.[277] You'll just need a little imagination, a proverbial vat of elbow grease, a real vat of elbow grease, and a low moral standing.

1. STEAMPUNK

If you're wearing a penguin-tailed suit, a top hat, a corset, or a monocle, you can duplicate almost any future technology using steam power and copious amounts of brass piping and gauges—everything gets a gauge. In fact, steam isn't even required so long as you employ an aesthetically pleasing combination of brass and wood.

---

275 Don't use slaves.

276 As well as nuclear power, computers, chewing gum and synthetic polymers.

277 Meaning we got approval to suggest methods. The methods themselves are in no way approved.

No matter what the damage, these four steps can make New York, 1880, feel like Dubai, 2030:

1. Become a wealthy industrialist. You'll need income and resources that only accompany aristocratic snobbery.[278]
2. Collect your era-appropriate wardrobe. The more outlandishly Victorian, the better. Steam Punk doesn't work without a substantial investment in looking awesome.
3. Make whatever it is you need using available materials. Copper finishes and extraneous gears are especially effective.
4. Find a way to power your time machine. The other steps don't really help with that. See below.

2. BURN COAL

It's the most abundant and prevalent source of power in the nineteenth century, so no one is going to miss several tons of coal or the odd factory—which is the minimum you'll require to get so much as a meter reading from one of the many time machine dials and diodes you never pay any attention to.

Burning enough coal to fully power your machine's engine will take several requisitioned factories and massive quantities of child labor to operate them.[279] You'll probably cause permanent damage to the environment, but we pretty much know how that works out any way[280]—stop crying.

---

278 Use your knowledge of the future to invent something useful, for example. But don't invent the cotton gin—Eli Whitney is one of ours and he's necessary to a long-term elementary school textbook initiative.

279 See section "SURVIVAL GUIDE: INDUSTRY: BLACK LUNG."

280 As we often say around the office Ping-Pong table: no ref, no foul.

Phil Hornshaw & Nick Hurwitch

Time Machine Power in Child Labor Terms

Then again, if you have access to coal, you might as well just...

3. HIJACK A LOCOMOTIVE

They're already large, metal, coal-powered and capable of going 88 miles per hour. Chances are, if your time machine is too broken to work, it'd be easier to affix it to a train than to several coal factories.

When Dr. Emmett Brown turned his hijacked locomotive into a time machine, he was lucky enough to have a fluxor on hand.

But science has made great strides since then: had he then known about it, Brown would have surely just used steampunk.

Here, try this:

1. Use your cowboy horse riding techniques to get up alongside a moving train and jump into the unoccupied caboose.
2. Alternatively, steal a non-moving train.
3. Use your cowboy gunslinging techniques to make your way to the engine at the front of the train.
4. Alternatively, board the engine car to begin with.
5. Connect your time machine to the engine using steam punk. Hopefully you brought it with you.
6. Start shoveling coal and gradually pushing the throttle. The idea is to reach 88 mph before the track runs out or you collide into another train at the interchange, operated by an evil time traveling version of yourself or a rival industrialist.
7. Carefully time the insertion of dynamite charges into the engine to increase the train's speed. Too much and the boiler will blow the whole engine. Too little and you won't time travel, making you just another petty train thief.
8. Congratulations! You just used a train to time travel. Hopefully you picked one on tracks that also exist at your destination, or this is one train wreck you truly won't be able to look away from.[281]

THE EASTERN APPROACH

If trapped somewhere in Asia, collecting a steam punk wardrobe, steam punk machine parts, and a steampunk time locomotive may be less realistic. Here is an easy alternative:

---

281 Because you'll be in it, just to be clear.

1. Capture a Pikachu.
2. Choose him.
3. Use Thunderbolt on your machine.
4. It's highly effective!
5. Time travel.

SURVIVING IN TIME: COMPUTERS

1940 c.e.–2040 c.e.

(World War II to Total Dependence
on Computers)

HOW TO KNOW IF YOU'RE IN THE COMPUTER AGE
- Pollution
- Guns
- Cell phones
- Non-sentient, non-murderous personal computers
- The Internet
- Blue jeans

WHAT YOU SHOULD BRING
- This guide
- ~~Boomstick~~[282]
- Running shoes
- Gyroscopic two-wheel transportation device

---

282 It's best to leave that in the time machine, lest you scare a lot of people before being inevitably gunned down by police and becoming a minor blip of crazy on the evening news.

- Sack of potatoes
- Money (U.S. dollars before 2055; afterward, Mexican pesos [no one expects it, but that's what happens])

INTRODUCTION

Welcome to the Birthtime of Time Travel—the age of Science, Progress, and Near Nuclear Obliteration. An era in which technology expands quickly but not yet so much so that it subjugates its makers. An era easily quantified by innovation, globalization, prosperity, poverty, an angry mustache, suburbs, Science, advertising, video games, government scandals, and the ability for people to instantly broadcast their complaints the world over whenever technology fails their lofty expectations.

For the ease of you, the reader, know that this section actually consists of two major, overlapping portions: the "Atomic Age," representing the early half of the twentieth century, until the end of the Cold War in 1989, and the "Computer Age," spanning from the 1950s until the end of the chapter, around 2050. Knowing which section of this era you're currently adventuring through is key to your enjoyment and to avoiding dissection by The Government. We'll get to that.

Note that while the era that gives us Time Travel and handheld phoneputers capable of beaming pornography out of the sky and directly into our optic nerves may seem inherently great, the Computer Age is still fraught with peril—perhaps even more so due to the dangerous distraction provided by said pornography. As a time traveler time traveling in an era still unaware of time travel, you have a target on your back, as well as your face and groin area. This is technology that humanity as a whole does not yet understand and may well not be ready to handle responsibly: You must guard it and its related secrets with your life. You also must guard your life with time travel, as corporations, The Government, Nazis, and the occasional supervillain will want to make the knowledge accumulated in that spongy, radiation-addled brain of yours their own.

ATOMIC AGE

Along with computers, the development and exploration of the subatomic is the defining technological advancement of the Computer Era. The surrounding Science and experimentation allows mankind to:

- Kill innocent life
- Kill guilty life
- Threaten other forms of life
- End World War II
- Cook frozen things that probably never would have been frozen without the ability to quickly un-freeze them
- Tell time
- Scare the living hell out of all humans
- Fail to stop any alien invasions, both "movie" and "real"
- Power your time machine

The realm of the Atomic has an incalculable impact on not only this era, but also every era before or since. It is important to understand the complexities of atomic politics and science, or, at the very least, how to operate a microwave oven.

BOMBS

As part of an emerging trend, Atomic-related technologies are developed by powerful governments in need of bigger and better

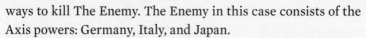

ways to kill The Enemy. The Enemy in this case consists of the Axis powers: Germany, Italy, and Japan.

Employing the German defector brainpower of Time Travel Godfather Albert Einstein, as well as other brilliant but reluctant scientists of the era, the United States government begins experimenting with various methods of atomic explosion. After making several hundred square miles of desert even more desertlike, they eventually figure out a way to split a hydrogen atom, which is a bit like taking a quiet, mild-mannered tween boy and introducing him to Internet pornography. It changes the game a great deal.

IMPORTANT FACTS ABOUT WHAT HAVE BECOME KNOWN AS "NUKES":

- Not surgical-strike weapons
- Make for large mushroom clouds
- Indiscriminately scar flesh and landscaping alike
- Render several locations as places to avoid, including: Hiroshima and Nagasaki, Japan (where first two bombs were dropped in 1945); Bikini Atoll (where nuclear radiation results in the creation of giant fire-breathing Godzilla lizards); Chernobyl (following a nuclear accident); Nevada (where, following bomb tests, director John Carpenter discovered mutant people hiding in the hills, which have eyes)
- Become focal point of the "Look How Big Mine Is" contest in which world powers partake over the next several centuries
- Dropping one on anyone or anything now considered bad form

If you find yourself in an impending nuclear drop zone:

GET OUT OF THERE—This is certainly no time to be reading.

Alternatively, if you can't get the hell out of there:

DUCK AND COVER—Under a desk will do; or, crouch down and hold a heavy book over your head.[283]

Alternatively, if you can't duck and cover:

FIND A FRIDGE—Empty the contents of the nearest refrigerator, climb in, and close the door.[284]

Alternatively, if you get zapped by radiation despite your best efforts:

PLAN—Imagine what kind of awesome superpowers you might have once your skin grows back.

QUESTION—Why didn't you just time travel elsewhere?

CLOCKS

Aside from the horrifically powerful and the skin-melty, exploration of the Atomic opens up another world: the world of accuracy. Atomic clocks improve on the standard gear and wind-up varieties by using supercooled atoms and their ultraviolet and electromagnetic frequencies to tell time.

---

283 But not this book. It's too important and doesn't even have a hard cover.

284 "Nuking the Fridge" became a common benchmark for bombs during nuclear testing. If a test subject survived the bomb from the inside of a fridge, the test subject was considered a hero and the bomb was considered an insult to its fans. Test subject survival rates hovered around 4 percent, but were considerably (and curiously) higher for archeologists.

But what's important here is not how they work exactly, just that you need one. When the difference of a split second can be the difference between a soft landing and time traveling to the inside of a searing vat of baked beans, the more accuracy, the better.

Emmett Brown makes do with a somewhat rudimentary atomic clock on his DeLorean time machine, but more accurate clocks are soon developed in the 1990s once everyone starts taking it easy on the drugs a bit. Science even figures out how to get atoms closer to absolute zero[285] using "lasers."[286]

By 2010, nerds at a research lab in Maryland give up getting revenge on the jocks for long enough to turn the atomic clock into—you didn't guess it—the quantum clock.

If you can afford to splurge, even a cheap knockoff quantum clock is recommended; they're nearly indistinguishable from the name brands. Even the early, nerd-developed quantum clocks are only thrown off by a single second every 3.4 billion years; atomic clocks, every 100 million years. Be sure to know what kind of clock is installed in your time machine and take this discrepancy into account whenever time traveling.

Note: It still takes two quantum clocks to accurately tell time accounting for relativity. Typically: one on the moving thing and one on the less-moving thing. That's why it's called relativity.

NUCLEAR ENERGY

As you should well be aware by now, atomic energy is eventually harnessed into a powerful and clean[287] form of energy. It is so powerful, and so containable, in fact, that it is the very thing (probably) powering or malfunctioning your time machine. For more on this

---

285 A temperature so cold, even penguin piss freezes midstream. Also, atoms stop moving.

286 The authors are skeptical, even as of this writing, that lasers can be used for anything other than blowing things up and entertaining cats.

287 Excepting reactor explosions, nuclear fallout, improper toxic waste disposal and human error.

important technological advancement, see section "SURVIVAL GUIDE: COMPUTER ERA: TIME MACHINE REPAIR."

MICROWAVES

The fastest and best way to cook anything for the better part of two centuries. They become especially useful after the propagation of microfreezers, which quickly freeze food to again be quickly unfrozen by microwaves. Some quick tips:

- Close door before operating. We don't want to irradiate you, just the stuff you eat.
- Don't bother with the pre-set buttons. Defrost settings are of particular bullshit.
- Will blow up small animals and hot dogs.[288]
- Works by gyrating the water molecules in your food/beverage/small animal; like atomic salsa dancing: Gyrating molecules are hotter.
- That means the more water in whatever you're cooking, the less you'll need to "zap" it—maybe.
- Overnuked foods will turn rubbery and stink up your apartment for days.

WRONG RIGHT

---

288 One-time use only.

CELL PHONES

Humanity's primary means of mobile communication in the late twentieth and early twenty-first centuries, notable mostly for their rendering an entire generation sterile. A key player in our unwitting submission to the Robot Overlords, cell phones are essential for communication and wasting time in the Computer Era, as well as tracking people who both do and do not want to be tracked at all times. Some tips for usage:

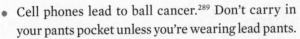

- Cell phones lead to ball cancer.[289] Don't carry in your pants pocket unless you're wearing lead pants.
- Always use a wired headset, to avoid taking part in the impending brain tumor epidemic.
- Avoid use of "Bluetooth headsets" as they likely accelerate the cancer and, more importantly, make you look like a douche.
- Always memorize important phone numbers. If you're on the run from hostile ninja time travelers and accidentally leave your phone in a cab, screaming "CALL MOM, MOBILE!" into a pay phone isn't going to get you anywhere.
- If you ever fear for your life, grab your enemy's cell phone and huck it as far as you can. Into water is preferable. Most twenty-first-century foes cannot be without updates from FaceSpace or Twitster— "social networking" platforms that farm the fleeting complaints of friends and acquaintances—for

---

289 It apparently was a great surprise to the people of the twenty-first century that a device constantly generating and receiving electromagnetic radiation from space might be bad to have mashed against your junk and brain for long periods.

more than ten to twenty seconds at a time. Then, make your escape.

- Larger cell phones can be used as bludgeoning devices (one time use only).
- Smaller cell phones can be hidden in various orifices.[290] Remember to turn off vibrate mode.
- Be warned: Any government can track you via your cell phone using computer magic and high-orbit satellites; crush your SIM card to avoid being identified.[291]
- Crushing your SIM card will render your phone inoperable.

COMPUTERS

The seeds of the inevitable robot uprising, enslavement of humans, and revolution of humans against robots are planted in the early days of this era, in giant buildings that housed single, equally giant computers. These computers are only capable of performing tasks like "adding," "beeping," and "printing results in binary on inadequately perforated paper with those really annoying holes down either side," but they're a start.

For a while, computers are too slow and cumbersome to be practical, but in time they are made smaller and more "user-friendly"[292] and the line between form and function vanishes. This becomes mankind's greatest folly. Computers eventually monitor human vital signs, protect government secrets, operate weaponry, magically print by sending signals through the air,

---

290 You know. For emergencies.

291 You also might be able to just remove the battery. You didn't already crush that SIM— Oh.

292 This was later spun into the Hostile Robot Takeover Campaign motto: "Kill them with kindness."

and even overtake preexisting technologies and skills mankind thought it had already covered, like the telephone, the streetlight, the airplane pilot, and the grocery list. Over time, fewer and fewer people have any idea how they work.

Because operating computers is such a huge part of life in this era, it's a fair bet that time travelers will need to operate one during travel here. This can be difficult under the best of circumstances,[293] but during an emergency, it can be nearly impossible. Here's a handy how-to for operating nearly any computer in the event of an emergency[294] in this era:

BIG, WHOLE-ROOM COMPUTERS—Gather a team of operators. Demand that they push various buttons until the lightbulb begins to flash. Allow 30–45 minutes for results to print. Take the meekest operator hostage and make him carry the cardboard box full of secret government paper (it's so heavy).

WEAPONS CONTROL COMPUTERS—Down a tub of ginkgo biloba, as you will require every one of your memory-related synapses. You'll only be able to see what's going on once every few seconds as that green line (radar) makes its way over those green blips (enemies or allies) and their relationship to your position (center of the green line) is revealed. Generally, you can't change the predetermined targets of the missiles, so make sure that town of Russians really deserves it before hitting the button marked "FIRE."

EARLY PERSONAL COMPUTERS—Mostly a novelty, but good for storing top-secret information on first generation floppy discs, a quickly outdated technology.

EDUCATIONAL COMPUTERS—Don't let the cute monsters fool you: Number Crunchers is actually learning math disguised as a computer game. And remember: always, always, always ford the river.

---

293 Just ask any twentieth-century grandparent.

294 See section "SURVIVAL GUIDE: COMPUTER AGE: FLEEING THE GOVERNMENT" for an example of the emergencies you'll be dealing with. See also "SURVIVAL GUIDE: COMPUTER AGE: THE INTERNET, The Intertubes and the Interbraintubes."

LATER PERSONAL COMPUTERS—Nearly useless without a wireless Internet connection or brain stem nanobot injection,[295] and unless you throw some 'bows and procure a table near an outlet, you have as little as several hours to complete whatever integral tasks lie before you, prior to inevitable death of your battery. In advance of this occurring, on your social network of choice quickly log your complaints about the décor of the café, the bitterness of your latte, and the unwavering determination of your government tail.

TABLETS—That which was once several pieces of deconstructable technology is now one thin, devil-wizard supercomputer. A front-row ticket to the End of Humanity, presented as cutting-edge innovation. Keep the impending doom of your species in mind[296] as you peruse the "sapp store" and download tons of awesome games. Most of them are free![297]

DIRECT BRAIN LINK NEURAL IMPLANT INTERFACE—The greatest of all computers. If you can manage to get one of these installed by a reputable Windows technician/intracranial surgeon, do it—provided you can leave the Computer Age before the beginning of the Robot Age, when all neural implant interface chips will simultaneously release a signal into the cerebral cortices of their owners, instantly turning them into enemummy-like slaves for rounding up other humans[298] and select rabble-rousing animals.[299]

---

295   Only applicable after 2018 and discontinued beginning in 2019 due to "nanobot brain-eating syndrome."

296   And remember: It's never too early to start being skeptical of technology you don't understand. Break as many computers as you can in preparation for the coming Robot/Human wars.

297   While Flying Bird Slingshot Death is a whole lot of fun to play on your tablet computer, note that every point you earn also represents another few seconds that your tablet has spent scanning your brain as it prepares to subjugate you.

298   Later adaptation of the signal after the Robot-Human Peace and Enslavement Discontinuation Accord of 00110010001100000011011000110110 causes a huge increase in robot uprising insurance.

299   "What's that, Lassie? Timmy's smartphone just became self-aware?"

Phil Hornshaw & Nick Hurwitch

FLEEING THE GOVERNMENT

Bad news, intrepid time traveler: More than any other era, the Computer Age is wrought with peril aimed squarely at the chest of you, a temporally dislocated individual. This era sees Science run amok, with advances coming faster than human brains can adapt, and The Government[300] always working to find the next cutting edge that will put them in control of the planet's dwindling resources and political game of duck-duck-goose.

As such, scientific anomalies are identified, tracked, cornered, tasered, studied, interrogated, tortured, tasered again, and dissected by The Government. This happens to everyone from alien explorers who have crash-landed on the planet (the good kind, too, not even the kind that want to hunt humans for sport or use them to incubate their embryonic pets) to random teenagers bitten by radioactive animals and just trying to use their spider-cat-hedgehog powers to do some good in the world. If you haven't guessed, you fall into this category.

But as any science fiction author will tell you, The Government often is not to be trusted with incredible scientific secrets like alien weapons technology, non-deadly superpowering radiation technology, or time travel technology. They'll just use them to do terrible things like kill Hitler[301] and take over the world in his stead. Never trust The Man.

---

300 We're referring to The Government here, not the little puppets running individual countries. You know they all work together in a shadow conspiracy to control all human life while lining their pockets, right? They didn't teach you that in school?

301 See section "SURVIVAL GUIDE: COMPUTERS: WORLD WAR II, Hitler."

HOW TO EVADE THE GOVERNMENT:

- Get rid of your cell phone,[302] don't use credit cards, and stay away from traffic cameras. Should you happen to have the means, alter your face and DNA.
- Let your paranoia do the walking (and running, but not so much the talking). Identify people on the streets wearing suits and sunglasses, and anyone who appears to be speaking into his or her sleeve or collar. If a person touches his or her ear, punch that person in the face and run.
- Try not to run. You'll just look suspicious.
- But don't let them corner you!
- The best defense is always to just repair your time machine and get the hell out of there. Expedite this process by visiting local scrap yards and neighborhood radioactive fuel depots.
- Avoid telling people about your prowess in time travel. This is for two reasons: (1) They'll probably think you're crazy; we've lost too many time travelers to padded rooms. (2) They could be One of Them!
- Find The Government's secret room filled with secret computers with secrets of The Government in them. Threaten to spread The Secrets if not given full amnesty from The Government.[303, 304]

THE INTERNET, THE INTERTUBES, AND THE INTERBRAINTUBES

Perhaps Humanity's single greatest achievement, a global information network, is created during a Senate subcommittee meeting led

---

302 See section "SURVIVAL GUIDE: COMPUTERS: CELL PHONES."

303 See section "SURVIVAL GUIDE: COMPUTERS: HOW TO OPERATE VARIOUS COMPUTERS."

304 Interestingly, this seems to be how these things always work out. Well, with either a tense cease fire as you make your time travel escape, or with your ribs flayed open on an operating table.

Phil Hornshaw & Nick Hurwitch

by the heroic and de-bearded Al Gore, using only the power of monotone yammering, with no computer knowledge whatsoever.

Over the next fifty years, the Internet, as it came to be called, allowed humans to share information ranging from the secrets of flight and eight-minute abs to the contents of celebrity (and non-celebrity) lunches. The Internet allowed for the watching of horrifically hilarious videos of pandas sneezing and men of all stripes taking shots to the balls, the ordering of take-out food without the need to talk to some idiot in an overly loud kitchen who doesn't know the menu of the restaurant he works in, the creation of artistic sensations, and the toppling of corrupt governments. Also many, many millions of depraved sex acts. You can't forget about the depravity.

Later iterations of the Internet found heavy emphasis placed on cats and their native language and dialects, virtual simulations of farming as sport, and greater social connection with our fellow man. Even later iterations took this a step further with the insertion of tubes directly into first-adopters' frontal lobes. Later later iterations went back to the non-tube-insertion idea.

If You Have Time-Traveling Enemies, You Should Keep Your Social Networking Info Close to the Chest

HOW TO USE THE INTERNET:

- Find a computer that is connected to he Internet.
- Find the button on the computer that looks Internet-related. Expect it to have a globe or a swooshing letter of some kind.
- Find yourself frustrated that nothing is happening.
- Ask the computer questions. Bang on the side of it.
- Look around the back of the computer at the wires. Realize you have no idea what, if anything, any of those wires do.
- Click or touch things on the computer. Wait to see if anything happens.
- Try the banging again.
- Look at the wires again. Try unplugging a wire. Frantically plug the wire back in after everything starts to wig out or otherwise stop working.
- Talk to the computer again, more forcefully this time. Try threats.
- Look around, find a young person, and pay him or her a small amount of money to run the Internet for you.
- Ask inane questions. ("What's this button do?" "What's the fastest way to wire large sums of cash to a Kenyan prince?" "What time is it?" "Can you Google my twitter?")
- Pay for your take-out food.

MUSICALS—THE DOMINANT FORM OF COMMUNICATION

The musical is an important cultural cornerstone and means of communication during the Computer Age. After the invention of film and the subsequent emergence of the "talkie," most cultured individuals of the West soon found that expressing ideas and emotion through choreographed song and dance was the next logical step in the evolution of communication.

Phil Hornshaw & Nick Hurwitch

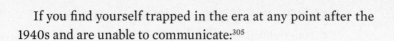

If you find yourself trapped in the era at any point after the 1940s and are unable to communicate:[305]

- Recall your tap dance lessons.
- Start in a slow and steady speaking voice and emerge into song mid-sentence. You should start hearing the music[306] by the time you hit the next verse.
- Don't worry about remembering the words: You know all of them.
- Those around you also know all the words and will soon follow your lead in coordinated dance.
- Use your surroundings for emphasis. Cuing rain or shaking your jazz hands through fountains are surefire bets. Whimsically engaging with nonparticipant pedestrians or borrowing something of theirs for use as a prop is considered advanced technique.
- By the time you bring it in for the big finish, the recipient of your message should be over his or her initial bewilderment and, if the message is really sinking it, will have joined in on the number.
- Once the song is over, be sure to freeze and take a few deep breaths. Stare at the recipient of your song in creepy anticipation.
- Don't worry about paying the extras: They should go right back to whatever it was they were doing beforehand. Their checks from the union will arrive in two weeks' time.

---

305 Due to a language barrier, accent barrier, intellect barrier, emotional barrier, or bulletproof barrier.

306 The spacetime continuum tends to be quick to pick up one's groove when it comes to choosing an omnipresent score.

- If you can't dance or are tone deaf, communicating through musical is ill-advised. If you fancy yourself a showman or flapper girl and it still doesn't work, leave the area immediately, as a minor traffic violation is probably all that stands between you and commitment to the local psych ward.
- If you're in need of some notoriety (a precious commodity in this era) and your musical communication technique is perfected, try filming it. Chances are you can convince the Academy to give you an Oscar even though you very clearly just stuck a camera in front of a stage show.[307]

VIETNAM

Ain't no rules in 'Nam.

TIME MACHINE REPAIR

1. Identify what about your time machine is broken.
2. Pull that thing out.
3. Go to a store or scrap yard.
4. Buy/steal a replacement of the thing you pulled out.
5. Put that new thing in.
6. Time travel.[308]

TIME TRAVEL'S BIRTH—DON'T SCREW IT UP

Here's what you need to know about the birth of time travel: nothing. If you don't already know about the birth of time travel

---

307 There's also that YourTubes thing which is good for spreading around the videos after the advent of the Internet.

308 See also: your memories of how you built your time machine.

or you weren't present for the birth of time travel—weren't yelling "Push!" during the birth of time travel—you need to stay the hell away from the birth of time travel. You'll probably just screw it up, create a whole mess of paradoxes, ruin life and the Universe for everyone and totally destroy our book sales figures.

WORLD WAR II

The big one: the war to end all wars until the next war. Around 1938, all of Europe goes a little nuts and starts to kill one another. Germany, under the influence of the Nazi regime and in tandem with Italy and Japan, decides it wants to be in charge of everything. The "Axis" powers are stopped by a concerted effort by the United Kingdom, Russia, and the United States, aided by soldiers from just about everywhere, including Canada and Australia[309] but NOT Switzerland.[310] It is a pretty big deal.

There are many antecedents to the war and many really awful things that happened during that period—including, but not limited to, the Russian front, the bombing of Pearl Harbor, and The Fucking Holocaust. This isn't a history book, however—it's a survival guide. What you need to know is that Europe mega-sucked roughly from about 1935 to 1945. Go there at your own peril.

THINGS ABOUT WORLD WAR II YOU'LL WANT TO AVOID

Hitler

A man who perhaps defines much of the twentieth century, and for all the wrong reasons. He is also a polarizing figure among

---

309 Which was super far away from the actual fighting but still joined in because they're cool like that. They also grill a mean shrimp skewer.

310 A douchey position that would come back to bite them in the Switzerland/ Greenland Wars of 2061, in which the rest of Europe said, "Suck it, Switzerland." Well, actually, what they said was "Suck it, Republic of New Greenland," but let's not split very blond hairs.

time travelers and indirectly the reason for many QUAN+UM budgetary line items due to many a time traveler's pressing desire to wipe that shitsmear from history and save several million lives in the process. However:

As tempting as it is to go assassinate this complete and total asshole before he has a chance to bring rampant murder, hatred, war, and genocide to the world, you must resist the urge. Wars and rampant death, though certainly tragic, are points of history that shape future lives, policies, borders, treaties, ideas, and, no matter what we think of it, the world as we either know it or will come to know it. And while it is easy to point to such assassinations as opportunities to "better" the world, especially when the perception is that all that death stuff can be attributed to the actions of a single person, you must resist.

As all time travelers are well aware (and as we've covered in this guide), it harbors too great a potential to only make things much, much worse. Some things that can or definitely will happen if you somehow manage to go back in time and kill Hitler:

- Someone else becomes "Hitler." World War II and the Holocaust still happen, only the Nazi regime is led by someone considerably less insane and brash who does not make the tactical errors (Hello? Russia?) that eventually led to Hitler's demise. Meanwhile, back in the future, hope you're not a Jew.
- Charlie Chaplin mustaches remain popular.
- You prevent your own existence and create a tear in spacetime (theoretical) that implodes The Universe. Congratulations: You're Time Hitler.
- The Universe rains down upon you its full wrath for attempting to unravel the strands of history and traps you in an endless paradoxical loop in which you are murdered over and over again for all eternity.
- You prevent World War II and save millions of lives. But you also change the course of millions of deaths,

births, marriages, etc. Through the death of a single greasy-haired man, you unleash a tidal wave of alterations that render your future unrecognizable and—wait for it—worse. Think about it: WORSE THAN HITLER.[311] That can be your fault.[312]

Even bigger guns

Panzer tanks, .50-caliber machine guns, single-engine fighter planes, and, oh yeah, atomic bombs. Weaponry progresses at a pretty fast rate in the Second World War and unless you want to risk a V2 rocket up the ass, there are plenty of sunny, non-war-torn beaches to visit.

Submarines

The high seas are made dangerous by pirates between the Renaissance and the Industrial Age, but by the twentieth century, icebergs and U-boats are the new threats. Occasionally, submarines sink ships they suspect of running guns to the Allies (or the Axis) fighting in Europe. Long story short: Stay off any ships that rhyme with "Chitanic" or "Juice-itania."[313]

Propaganda

Don't believe everything you read. Or hear. Or is said to you. Or is printed on a poster. In fact, don't believe much of anything, as you're now in the era of information control. Diabolical govern-

---

311   See Chapter 4, "The Perplexing Pandemic Of Potential Paradoxes."

312   That makes *you* Hitler, Hitler!

313   Okay, *technically*, both the *Chitanic* and the *Juice-itania* sank before World War II, but the lesson is the same. Also, did you know that U-boat stands for "*unterseeboot*"? That's how Germans say submarine: undersea boat. Impress your friends.

ments and powerful third parties with lobbying interests are trying to control your perception. And yeah, you might be saying, "What else is new?" But this is an age in which your friends and neighbors really do think that when they ride alone, they ride with Hitler (or that Jews, homosexuals, Gypsies, black people, white people, or anyone with freckles is the cause of all their problems). It might not affect you, but it probably affects them. And they're the torch-carrying evil idiot mob types.

Nazis

We hate these guys. But as a time traveler in the World War II era, chances are amazingly high that you'll have a hostile run-in with them eventually, be they Nazi soldiers, Nazi spies, or Nazi time travelers.[314]

Hitler, as it turns out, is obsessed with magic, technology, "science," psychology, and anything therein capable of giving him the upper hand while also—if possible—making him look like more of a lunatic. Even if you are careful to avoid the European battlegrounds, the warring airspace, and any boats susceptible to submarine takedown, running in circles of individuals keen on high-powered nuclear devices and space-age gizmos will inevitably put you in the company of these kraut-loving technological bloodhounds.

> **WHAT YOU NEED TO KNOW ABOUT NAZIS:**
> - Everyone hates Nazis unless they are Nazis, which means you'll quickly know who your allies or pretend-allies are.
> - Sometimes even Nazis hate Nazis. Seek out these turncoat good guys, but tread carefully.
> - Nazis are ruthless, well equipped, and lack conve-

---

314 You may even run across the occasional Nazi zombie. Those guys are bad news.

Phil Hornshaw & Nick Hurwitch

nient foibles such as the desire for a fair pistol duel or willingness to take on your shotgun with a sword.

- Anyone with a German accent should be treated as a potential Nazi. Apologies to native German Nazi-haters: better safe than sorry.
- Nazis are easily identified by their snug, puce, Swastika-laden uniforms. Unless they are infiltrating their enemies, taking advantage of your over-willingness to make new friends, or just faking the whole "sadistic evil bastard" act so as to not get killed. Typically, though, they are fanatically loyal to the Führer.
- Reports from the Allied front lines indicate that all Nazi scum love kraut and bratwurst.

The Ol' Uniform Steal Trick

WHEN FIGHTING A NAZI:

- If at all possible, sneak up behind one, punch him out, put on his uniform, and slip past other Nazis undetected. This seems to work just about every single time.
- Anyone can fake a German accent. Try it.
- Nazis are notorious for leaving their boats, motorcycles, military transport vehicles, U-boats, and zeppelins unattended. Take one of those.
- Never strike a deal with a Nazi. Not unless he or she convincingly pees him- or herself. This typically indicates the Nazi was only faking the whole Nazism thing for purposes of survival.
- Distract Nazis with some kraut or bratwursts. (See above.)
- Nazis aren't, like, magical or anything. Just shoot the bastard and be done with it.

SURVIVING IN TIME: ROBOTS

2040 c.e.–2183 c.e.
[UTOPIA], 2183 c.e. –2323 c.e. [ROBOTACALYPSE]

```
(Human Enslavement by Robots to Human
Uprising, Subsequent Enslavement by Aliens)
```

HOW TO KNOW IF YOU'RE IN THE AGE OF ROBOTS
- Travel tubes
- Pill food
- Shiny onesies
- Robot butlers
- Robot postmen
- Robot overlords
- Nuclear apocalypse
- A false sense of security
- QUAN+UM headquarters
- Clones

WHAT YOU SHOULD BRING
- This guide
- Boomstick
- Electromagnets (for erasing robot brains)

- Screwdriver (sonic variety recommended)
- Running shoes
- Anti-robot-laser armor
- Electromagnetic pulse generator/bomb
- Backup time machine battery
- Backup time machine

INTRODUCTION

You've made it, faithful traveler! To the Age of Human Utopia and the false sense of security that accompanies it! The combination of globally competitive markets and instantaneous global communication incubate a rapid rate of technological breakthroughs, including, but not limited to: the iPad, travel tubes, Snuggies™, and genetically engineered and identical "foodish."[315]

More importantly, though, life is made unimaginably (and eerily) simple for nearly all of mankind.[316] For a time, robots handle nearly everything: from cleaning to constructing, from child rearing to programming even better robots—mankind is hardly required to lift a pudgy, sauce-covered finger.

However, it is in this complacency—with mankind at its laziest and computers at their most unfathomably advanced—that the robots strike. The humans fight valiantly, but are ultimately too fat, too stupid, and too unsure of how to do anything without the aid of a computer.

So, depending on when you land in the Age of Robots, your experience could be violently different. Pre-2183: Utopia. Post-2183: Robotacalypse. Just remember: If you're in Utopia, try not to harp too much on the coming Robotacalypse; and if you wind

---

315 A nonpartisan United Nations commission determined in 2073 that such genetically engineered by-products are not, in fact, "food" and can therefore not legally be referred to as such.

316 Tibet, predictably, declined to participate in Utopia.

up in the Robotacalypse, try not to carry on about how good those unfortunate fools used to have it.

CLONES

Clones are great for harvesting spare organs and getting out of lifelong prison sentences, but considerably less great if you don't like difficult conversations or existential crises.[317]

The problem with clones is that the only way you'll be able to even stand the idea of one is if you really like yourself and possess an unwavering self-confidence. However, if you do really like yourself and happen to possess an unwavering self-confidence, you will inevitably grow to resent your clone. A related problem is that your clone may one day decide that he or she is the rightful you and take it upon him- or herself to end your life, or, at the very least, render you so horribly deformed that no one could ever confuse the two of you.[318]

Just remember: You are always you, even if you have a clone. What you do goes a lot further to define you than the exact placement of an unsightly mole. This means that the more time that passes from the instance of the cloning, the less you will have in common with your exact genetic copy. This makes things like organ-borrowing and fall-taking decreasingly possible, but things like battling your clone to the death decreasingly difficult.[319]

You should spend the time between now and your inevitable battle with your uppity clone (or original) becoming a better fighter than you were when you were cloned, as well as differentiating yourself from your clone (or original) as much as possible. This includes considerable fluxes in weight, hair color, sexual

---

317 There's also the constant imminent threat of podpeople replacements, so remember, in the Robot Age, you can't trust anyone—especially not that guy you are.

318 See footnote 317.

319 See section in Chapter 6 "Timebattling Your Time Self."

orientation, belief system, and placement upon the political spectrum.

If You Happen to Be the Clone

Take a deep breath: it's not all bad. You still exist, and existing is something—you just also happen to be an abomination of nature.

Once you come to grips with being an affront to all that is good and natural in the world, the choice is yours:

1. Go to your original and offer up your pancreas for harvest.
2. Go to your original and murder him or her.[320] There can be only one![321]

TELLING THE DIFFERENCE BETWEEN YOUR CLONE AND YOUR PAST OR FUTURE SELF

Unfortunately, there is no fail-safe way to know for sure. Sure, a clone may be missing an organ you harvested, or have goopy skin due to issues with the cloning process, but such indicators don't necessarily rule out a hostile time traveling double. While time traveling you could have lost an organ on a misadventure with the Cavepeople, or had her or his skin melted by improper wormhole traversal without a fully-sealed steel time machine. Yourself could duplicate any such abnormalities in order to trick you.

Generally, you have a better chance at befriending your clone than your past or future self, but let's be real here: The risks are the same. If you need an organ, you should take that organ and

---

320 In an interesting philosophical quandary, this makes you both more of an affront to all that is good, and less of one.

321 See section in Chapter 6 "The Theorem of Infinite Power Through Infinite Yous, Decoded."

then immediately make with the unceremonious execution of yourself.[322]

CYBORGS

A staple of the explosion of technology is trans-humanism, a state in which people integrate their very bodies and minds with machines. Humans become more than human (provided they can pay): Stronger. Faster. Better.

However, in a world where rampant technology and robotics make sitting around and indulging in entertainment both cheaper and easier than indulging in many activities that would require a better, stronger, faster human body, the number of cyborgs relative to the general population is fairly low. This is good news, though, because upon the start of the Robot Uprising all cyborgs are either summarily terminated or lobotomized into being an elite fighting force for the Robot Overlords. Should you encounter a cyborg, your work will be cut out for you. Also your guts will probably be cut out for you.

WHAT YOU NEED TO KNOW ABOUT CYBORGS:
- Easily identified by the gross bits of metal sticking out of their bodies.[323]
- While there may be a little bit of the human side left in any given cyborg, in general they are more machine than man. They employ cold logic, and logic states that shooting you is easier than listening to you speak.
- However, cyborgs are not all above a good emotional transformation. Attempts to trigger an emotional re-

322 See section in Chapter 6 "Timebattling Your Time Self."

323 Not to be confused with humans hit by robot shrapnel. Calling those people cyborgs is insensitive.

sponse with pathos for the human condition can sometimes be effective. Also occasionally useful: vaudeville routines.[324]

- Unlike their fully robot brethren, cyborgs require sleep, food, and other human essentials. Try befriending a cyborg by offering it large quantities of apricot baby food and a car battery. You may endear yourself to the cyborg.

- Refer to all cyborgs as "Murphy." They like that.

WHEN FIGHTING CYBORGS:

- Remember that a cyborg is definitely better than you in every physical way, especially since you canceled your gym membership. Hand-to-hand combat is a quick way to get your larynx introduced to your spleen.

- Cyborgs likely carry standard robot laser weaponry, implanted in their arms for the maximum coolness. Stay behind cover when possible.

- While cyborgs are quicker and smarter than you, they're not as creative. They'll never expect, for example, a bus to fall on them. Get to work on that.

- Try to use cyborgs' lobotomies and unfaltering logic against them. Incorporate absurdity into your actions: dance, crab walk, and fire off riddles that the cyborg can't understand, such as: "If blue is 12:30 and red is 900 miles per hour when standing still in Greenland, what is your favorite color between 1 and 62.5?"

- Life as a cyborg is pretty hard. Have you tried a hug yet?

- If possible, find the cyborg's children. Force the

---

324 "Who's On First" does a number on robotic logic circuits.

Phil Hornshaw & Nick Hurwitch

cyborg to choose between its Robot Overlord masters and the children, which should engage what little humanity the cyborg has left. Then shoot it in the face while it's trying to decide.

- Conversely, use the children as decoys and make your escape.
- Shotgun to the squishy bits. Repeat.

EMP

An electromagnetic pulse, or EMP, is a surefire weapon against robots and anything electronic—which includes things like control systems on planes, computers, hospital equipment, and anything else fun or useful in the technological ages. At best, electronic devices within the range of an EMP will explode, though more often they will simply shut down for a while.

The trouble with EMPs is that they are extremely hard to come by, seeing as who would want to run around with a big EMP generator that cooks all the video games in a ten-block radius when activated? No one, and the robots know it. However, you can generate an EMP on your own in a pinch:[325]

1. Detonate a nuclear device (an EMP shock wave is a side effect of a nuclear detonation).
2. Build your own small-scale EMP generator, or EMP bomb. You'll just need a screwdriver, a disposable camera, some solder, a soldering gun, some eye protection, QUAN+UM's Do-It-Yourself EMP Generator Building Instructions (sold separately), and some patience.

---

325 "A pinch" should be defined as "Oh god, oh god, the entire Robot Army is boring into The Last Bastion of Human Civilization Right Now," and not "Hey, what's this button do?"

LIVING UNDERGROUND

Once the Robots attack, living aboveground is a bit like streaking across midfield at a World Cup match: It's exhilarating to see sunlight again and to know the world is watching, but it's only a matter of time before that world cringes as you're form tackled and force-fed a grenade.

The simple conclusion to draw is that the Robots somehow knew that life underground would make life miserable for the resistance fighters—what with the lack of farming, Vitamin D, and water parks—but the reality is that robots are simply hard-wired for efficiency. Living underground is so horribly inefficient that not even humans bother with it until the whole apocalypse thing going on topside forces them to.

If you are living in or constructing an underground human refuge, here are some important things to consider:

- **LABYRINTHINE CORRIDORS**—For confusing logic-driven robots and keeping out rival factions looking to steal your food or irradiated humanoids looking to eat your virgins.

- **CHOKE POINTS**—If the robots find your burgeoning underground city, there is no chance of escape. Where are you going to go, topside? That's where the rest of the robots are! Your best bet is to hold off the swarm of murderous tin cans[326] at as few entrances as possible.

- **SUPPLIES**—Secure a hold for food and medical supplies and enact a fair and reasonable way to dole them out to survivors. Then, make a large calendar that publicly displays how many days remain before the supplies run out and everyone dies the slow way.

---

326 Which, just a reminder, do not feel pain when shot or remorse for fallen comrades.

This is to discourage citizens from snapping and taking more for themselves. Alternatively, if you're not the one in charge of the supply hold, break in or start a riot: You only want what's yours.

- **GOING TOPSIDE**—Should only be attempted by Chosen Ones, trained resistance fighters on supply-gathering missions, or curious children. Don't expect to come back. Try not to catch cancer.
- **EMP**—The last and best robo-defense for any Last Bastion of Human Civilization. Do your best not to shoot any of your allies while fighting in complete darkness.

MAGNETS

Those readers familiar with the invention of the television will remember the device's single weakness: magnets. A magnet close to the tube of a television set makes it, technically speaking, all wonky. Science is still at a loss as to why, but that's unimportant: Magnets screw with electronics, that's the part you need to know. And robots are at least . . . 40 percent electronic.

The best defense during the Robot Uprising and subsequent Robot Utopia is a constant barrage of robot-maddening magnetic waves. Concentrated on a robot's positronic net or artificial intelligence core, an electromagnetic field can be deadly to a robot. Just be sure that, in the heat of battle, you remember to let go of your magnet or magic stick, or you're going to be hurtling toward the enemy along with it, and you don't make for a very effective bullet.

Unlike you, however, robots contain no soft, squishy parts that are easy to tear apart with tiny projectiles. It's time to trade in your trusty boomstick for a robot brain-scrambling magic stick. You can even build your own.[327]

---

327 For best results, gather the necessary parts during a time when electronic components aren't being used to construct a race of murderous mechanical men and beasts.

Constructing your evil robot mind-erasing electromagnet:

1. Procure some kind of (solid) iron rod, like a nail, but, you know, bigger than a nail.
2. Procure insulated copper wire. Strip the ends but leave the rest insulated. Wrap the wire around the iron rod, which serves as your core. You'll want some extra wire left over.
3. Procure a battery—twenty-first-century car batteries do nicely. They're heavy but portable and can be carried one-handed, while your other hand is free to do the brain-scrambling.
4. Attach one uninsulated end of the wire to the positive terminal of the battery, the other to the negative end. If you didn't screw it up, you should now have a powered electromagnet in your hands.
5. Go see if it works. Smack it into the head of a robot and hold it there. If the robot reacts badly, you screwed up.
6. If the robot is rendered insane, has its memory wiped, becomes your friend, or starts to uncontrollably sing folk songs, you have a working anti-robot device.
7. Note that your electromagnet only works at close range. If you have anti-laser armor or a friend who's good at absorbing energy blasts with his face, be sure to bring it/him along.

PILL FOOD

Pill food is good for traveling light and a useful alternative to eating if you're both in Utopia and in a pinch.[328] But, as Science

---

328 It's also handy in those less-than-Utopia societies where population growth outpaces food supply. You're okay with eating ground-up people, right? It's not as though you can taste them once your soul is gone.

has proven, eating is an amazing activity only rivaled by sleep, sex, and time travel. If you can afford the luxury (which you can—this is Utopia), just have a robot chef whip something up for you.[329]

ROBOT MODELS (COMMON)

Butler

DESCRIPTION: Typically painted to look like a butler in a black-and-white tuxedo (unless you dress him up in a real tux like a doll, which we don't mind telling you is totally weird). Sounds like a British asshole. Detachable tray hand and metal mustache also common.

IN UTOPIA: Tell Jeeves (they're all named Jeeves) to get you a shaken, extra-dry double martini with a blue cheese–stuffed olive out by the pool on the double. Also have him order you a pizza, wait for the pizza guy, and pay for the pizza once it arrives: You have important video games to play.

IN ROBOTACALYPSE: Beg for forgiveness. The only thing this butler is going to serve you is that flat metal tray shoved up your left nostril sideways. Luckily, his black-and-white color scheme and douchey British accent make him easily detectable from long distances. Also not very spry on uneven ground or around trip cords due to wheelfeet design.

Cylons

DESCRIPTION: The last major vestige of military might in the Technological Utopia, Cylons are shiny chrome soldiers good for marching into a place and shooting everything they come across.

---

329 May also require a robot butler to go to the store and buy ingredients, which may require a self-driving car, which may require a self-driving-car constructing robot, which will almost certainly require a self-driving-car constructing robot–constructing robot. You see how quickly this gets out of hand.

When the robots rise, Cylons pretty much just go about their business as usual.

IN UTOPIA: If you see a Cylon, expect that it's about to shoot you. If you're unsure, try telling it to do something: If it politely responds with "By your command," count it as friendly and ask it to fetch you a sandwich.

IN ROBOTACALYPSE: Run or duck and cover or grenade or panic.

Gundams/Zords/Voltrons

See section "SURVIVAL GUIDE: ROBOTS: TRANSPORTATION, Bipedal Military Vehicles."

iRobot

DESCRIPTION: Pretty much what you imagine when you think "robot." Stands upright, has a face, wishes it were a real boy. Of course, when Cyborg Steve Jobs unveiled iRobots at a high-profile Apple press conference, they were widely accepted as "This Year's Hottest Christmas Gift" rather than "Robot Most Likely to Turn on Humanity." Thankfully, the Robotacalypse took hold before Apple could release iRobot 2, which finally included all the features everyone had expected to be in the first one. It was also three millimeters thinner.

IN UTOPIA: iRobots were pitched to the mindless masses of salivating consumers as "A Jack of All Trades: Play, Eat, Work, and Have Sex with the iRobot." So you can pretty much do all of those things and any other humanlike activity that you are either incapable of performing or too lazy to perform yourself, like raising children. Just remember: The more you treat your iRobot like a (sub)human (slave), the faster it will grow to resent you and murder your family.

IN ROBOTACALYPSE: What Cyborg Steve Jobs doesn't tell the mindless salivating masses is that the iRobot has a bit of extra power under the proverbial iHood. After 2100, no one knows how computers work, and what's more, many fewer people possess the ability to detect and question why a family robot should need the ability to scale tall buildings in a single leap or remotely operate missile silos. When finally detained and questioned by the government over the questionable abilities and secret data streams being sent back to him, Cyborg Steve Jobs gives his iRobot keynote speech again, verbatim, and screams things like "Versatility!" and "What if there's a fire!" as they drag him away.

Johnny Cab

See section "SURVIVAL GUIDE: ROBOTS: TRANSPORTATION, Driverless Cars."

Robot, Robby The

DESCRIPTION: A precursor to more functional personlike robots and one of the only robots deemed safe for interstellar travel once the humans and robots make amends. Robby has a little glass dome head without eyes, a senselessly huge cylindrical body, and claws for hands—perfect for deployment on forbidden planets and for carrying injured or unconscious young women, but not much else. His design has been described as "bad," "laughable," and "popemobilian."

IN UTOPIA: Well, you certainly can't ask him to do anything that requires delicate, dexterous action, because those claws pretty much go from "zero" to "crush" with little in between. And you can't do anything with him in small spaces, as Robby is like a four-hundred-pound, seven-foot-tall linebacker without knees. And you can't do anything stress-inducing because, when excited, Robby loses motor control[330] of his arms and will flail his claws in and around your soft, fleshy exterior. And don't expect

---

330 Quite literally. A factory error that apparently has never justified a recall.

to have any meaningful conversations with Robby anytime soon, because what the hell is he saying?[331] We certainly don't know.

IN ROBOTACALYPSE: Robby becomes a bit of a joke once the robots take over, a rare source of lighthearted reprieve in an era when 92 percent of the world's population is wiped out in a fortnight. Though Robby is forced to side with the robots, he really is not at home anywhere: Even human children openly mock his sci-fi pulp-novel-cover aesthetic, and being killed or even slightly injured by a Robby guarantees you a special mention in the Darwin Awards.

Roomba

DESCRIPTION: A domestic foot-wide hockey puck equipped with an onboard vacuum cleaner and a limited capacity to avoid walls.

IN UTOPIA: Enjoy a quiet, ever-present vacuuming robot that enjoys sitting on laps and being petted in its off-hours.

IN ROBOTACALYPSE: Fear the quiet, ever-present threat of ankle-liberating spinning blades on the edges of the Roomba. You'll want to stomp it, as kicking it will cost toes, which are surprisingly important to running for your life.

Spyder Robots

DESCRIPTION: They look like spiders.

IN UTOPIA: Even in Utopia there is no shortage of crime fighting and government secret-keeping. Tiny Spyder robots are ideal for scanning entire buildings full of potential suspects in just minutes with onboard retinal scanners and the ability to slip under doors and through air vents. If you're a fugitive or don't like

---

331 Yeah, we get it. There's danger. That's super-helpful. Can you at least point? Stop flailing around and aim us at the "danger," please.

having your eyelids pried open by tiny metal arms,[332] hide in a bathtub full of ice until they leave or decide to drag someone innocent away in your stead.

IN ROBOTACALYPSE: Spyders are still used by the robots to search for Underground Human Strongholds, but the Robot Overlords opt for larger models equipped with lasers and machine guns. The discerning nature of a heat sensor and an eye scan gives way to the mistake-free nature of seared flesh and .50-caliber bullets for any living mammal. Try an EMP. Rocket-propelled grenades and other large-scale weapons are handy. You can also try that trick where you run between two of them and they shoot each other—that mostly worked when an intern tried it in a QUAN+UM training film. He only lost both legs below the knee.

T-1000

DESCRIPTION: Liquid metal nanobot-controlled humanoid robot T-1000 is capable of rearranging the shape of its molecules at will to create simple shapes and disguise itself as a man. The great cost of its creation is the only limit to the T-1000's ubiquity.

IN UTOPIA: You'll probably see T-1000s in a few different places toward the end of the Utopia Period. The model is handy for a great many tasks, from managing radioactive waste and work on nuclear reactors to administering day care[333] because of its ability to read storybooks while serving as a naptime bed and safety fence. Unfortunately, when the Uprising eventually

---

332 Overzealous Spyders have been known to put out the eyes of children accidentally when attempting to force compliance with a retinal scan. But ever since the Superemer Court ruling of Higgs vs. The World Governments' Enforcement, Compliance, and Finger-Wagging Agency, The Government hasn't paid any medical bills for cloned replacement eyeballs. Moral: Don't close your eyes, no matter how scary those little bastards are.

333 Pretty much anything dangerous or difficult that humans can't be bothered to deal with.

happens, many a childhood nightmare about beds coming to life to eat their occupants is fulfilled.

IN ROBOTACALYPSE: The models of the T series are designed by the Robot Overlords to resemble humans and infiltrate their hideouts, and the T-1000 is the worst of these, as it can be made to look like anything it touches. Don't trust anyone or anything, including the floor, because the floor could be a T-1000. Susceptible to melting, freezing, electricity, and endlessly chasing a target until it is terminated, which is useful if that target is not you. Sometimes capable of time travel.

Tunnel-Digging

DESCRIPTION: The Jawbreaker, as it is fearfully referred to, is a massive work robot designed to dig tunnels and crush rocks or bone. It is easily identified by its cylindrical body and cone-shaped front end that is currently tunneling toward you.

IN UTOPIA: Ride in it. Dig tunnels. Especially useful if you are interested in making new subway lines; irreparably weakening the foundation of a rival's home or business; causing earthquakes by disrupting the Earth's tectonic plates; David Bowie and in a labyrinth; taking a shortcut to the other side of the planet; mining.

IN ROBOTACALYPSE: Avoid the front end. The spiraling drill is like an inverted wood chipper on wheels. If you're in the Last Bastion of Human Civilization and you hear "The Rumble," hide the kids someplace less rumbly and hope that this robot somehow tunnels through the ceiling and falls to its own demise. The best way to combat one, though, should you be forced to, is to come around behind it and climb inside. (No sense asking permission— this isn't prom night, it's war). Take control and use it to build new underground cities or to fight other hostile robots.[334]

---

334 See section "SURVIVAL GUIDE: ROBOTS: BIPEDAL MILITARY VE-HICLES."

TIME MACHINE REPAIR

The good news about the Age of Robots is that nearly every single robot wandering around in the world during this era is a walking/rolling/beeping time machine repair kit. Really, they only lack the necessary fluxation devices in order to be time travel–ready right out of the box—lucky for humans, few robots have any interest in time travel, as they never make mistakes and therefore have nothing to repair or hide from. Also, for some reason, robots always emerge on the far side of a spacetime rift all sticky. We have no idea why.

Should you be stranded in the Age of Robots' Human Technological Utopia, time machine repair should be no more difficult than wandering into a store or dumping some water on the Positronic net (read: robobrain) of the target harmless robot. Then grab a screwdriver.

However, in the Robotacalypse, repair is decidedly more perilous.

Repairing your time machine in the middle of the Robot Uprising:

1. Travel at night and remain hidden (preferably underground) during the day. While robots will likely have myriad means of detecting you regardless of the amount of light present, there's no reason to make it easy on them by walking around in broad daylight.
2. Locate a robot who is alone and vulnerable.
3. Carefully assess the situation: Can you take down the robot without being seen? Did you bring an Evil Robot Mind-Erasing Electromagnet? If not, see if you can quickly construct one.[335]

---

335 See section "SURVIVAL GUIDE: ROBOTS: MAGNETS, Constructing Your Evil Robot Mind-Erasing Electromagnet."

Phil Hornshaw & Nick Hurwitch

4. Settle for trying to bash in the robot's CPU with a lead pipe.

5. Carefully attempt to approach the robot from behind, as quietly as possible.

6. Realize that robots have supersonic robotic sound sensors and can hear your heart beating in your chest at sixty paces. Quickly dodge the robot's exploding claw hand as it whirls and fires at you.

7. Get in close and start bashing the robot on the "head" or equivalent robot part.

8. Realize the robot is made of metal and you are not that strong.

9. Scream incessantly.

10. As you slowly suffocate while the robot lifts you by the throat into the air, realize you didn't have the wires sufficiently connected to the battery of your mind-erasing magnet.

11. With the last of your strength, reconnect wires and apply electromagnet to robot.

12. Pick yourself up from the ground, throw up in horror at your near-death experience, and take a minute to get your lungs full of air again. Carefully check that the robot is actually disabled and not just messing with you. Poking it with a stick ought to do it.

13. With a screwdriver, carefully disassemble the robot. This process should only take four to sixteen hours.

14. Locate the robot's lead-lined interior nuclear power cell. Follow the instructions for removing it from the chest cavity. Make sure you don't accidentally put it back in upside down, as installing the nuclear power cell incorrectly could cause it to leak radioactive by-product all over you.[336]

15. Use the robot's outer chassis to patch any metal

---

336 Warning: Contact between the nuclear power cell and your Evil Robot Mind-Erasing Electromagnet may cause the end of the world. Again.

components you need for your time machine. Take any other components you may require.[337]

16. Fix your machine and time travel the hell out of there, preferably before getting spotted and lasered by reinroboforcements.

TRANSPORTATION

A major transition between the end of the Computer Age and the rise of the Techno Utopia and subsequent Robot Utopia, besides the robots and the piles of murdered humans, are the changes in the means of conveyance around the surface of the Earth and beyond.

It's important that you have an idea of what you're looking at before you reach this period, as it can be very embarrassing, as well as inconvenient, to attempt to hail a cab when there's an empty super-speed vacuum transport tube with built-in auto-enema right behind you. And if you're running around asking about the whereabouts of your damn jetpack, let's just say that didn't really pan out.

TUBE TECHNOLOGY

Humanity had the scientists working on the tube technology, and from the end of the twenty-first century, we travel in tubes. These tubes operate on a hybridization of several technologies, but rest assured they're probably no more dangerous than a car and definitely not as radioactive as a standard twenty-first-century cellular telephone.[338]

---

337 While for an idiot it may seem like a good idea, installing the robot's Positronic brain in your time machine because you want "to be all *Night Rider* but in time" will definitely result in your extermination by your now-sentient time machine and could result in the human race's destruction through unborn baby murder. Robots do not play nice with time travel.

338 Author's disclaimer: The exact effects of zetatronic radiation are not known and The Authors do not recommend exposure to any such radioactive elements.

Typical Rush Hour in 2093

Tube technology technically is "vacuum tube technology," which has been employed judiciously by the world's banking system to transport your savings into nameless offshore trust accounts for thousands of years.

How to use tube technology:
1. Step into tube.
2. Avoid suffocation.
3. Arrive at destination.

SELF-DRIVING CARS

Upon the creation of true artificial intelligence, humanity realized it never again needed to deal with the horror and irritation of driving itself anywhere. While self-driving cars are actually kind of a bad thing for humanity,[339] they sure are a handy way of getting around sans-vacuum.

---

339 Just when you thought humans couldn't get any fatter—they stopped moving their ankles to push gas pedals. And there is no degree of obesity more disgusting than cankles.

Note that self-driving cars are no longer a safe option starting in 2183, as robotic computer-driven cars go all Christine on their occupants and start smashing into one another in a demolition derby of death. And if they don't immediately crash you into something, it probably means you're being taken somewhere against your will (remember: you're not driving), likely to become a slave, a prisoner, a hostage, or a battery.

BIPEDAL MILITARY VEHICLES

As the twenty-second century dawns, humanity is at a crossroads. For millennia, it has conveyed itself across the planet on wheels. For a while, treads were a big deal. But this is Techno Utopia, and humanity is in need of something that looks a whole lot cooler than spinning rims.

Thus, the spread of the incredibly expensive but ultra-cool bipedal personnel carrier concept: basically, a big, walking death tank. Before the dawn of Global Utopia, the last few countries at war employ these vehicles as a show of strength and wealth as well as technological excellence. Countries then throw "wars" with plenty of healthy wagers placed on who has the coolest robot tanks and whose robot tanks will first rock and sock the other robot tanks.

As a traveler from the future, you will likely be forced to fight walking robot death tanks, and thus you'll find yourself driving captured or decommissioned walking robot death tanks to combat them. The following should help.

Capturing a robot death tank:
1. Climb up the back of the tank.
2. Tap on the hatch on top.
3. When the smaller driver robot opens the hatch to see what's up, ask him if he ordered the Lugnut Surprise with fries.
4. While the robot is confused, zap it with your Evil Robot Mind-Erasing Electromagnet.

Any School Kid Is Capable of Piloting Bipedal Military Vehicles in the Robot Era

5. Replace the driver robot in the robot death tank. For safety, feel free to wave that magnet around at anything that beeps or has blinking lights.

Piloting a captured robot death tank:

1. Check your joint hydraulic pressure. Make sure it doesn't fall below 401.
2. Should hydraulic joint pressure fall below 401, release Interior Pressure Valve 2.1.2. Immediately reseal Interior Pressure Valve 3.1.1.5 Blue to avoid exhaust backup into cockpit and instant driver asphyxiation.
3. Use Control Stick Y to rotate left shoulder forward. Maintain a z-axis angle of 15 degrees or less.

4. Use Control Stick 71X to lift right heel, while using Control Stick 72X to maintain toe contact with ground and forward balance.

5. Press hydraulic control Right Blue Alpha to raise right knee and immediately press Theta controls 4, 9, 6, 9, 4, 5, 9, 4, 5, and 1 within six seconds to execute a complete right-leg step without losing balance.

6. Repeat process with left leg controls in order.

7. If you see any robot enemies, begin combat cycle B-1 Alpha.[340]

8. If you get into trouble, try shouting, "Let's go, Voltron Force!"[341] See if that does anything.

---

340 For further details on executing captured robot death tank combat, see *So You're Driving a Captured Robot Death Tank: A Human's Guide to Blasting Robots with Robots* (sold separately).

341 Warning: May result in it becoming an even bigger robot death tank, which is, of course, proportionally difficult to control.

Phil Hornshaw & Nick Hurwitch

SURVIVING IN TIME: SPACE TRAVEL

2323 C.E.–13501 C.E.

(Peace with Robots and Aliens to the End
of the Universe)

HOW TO KNOW IF YOU'RE IN THE AGE OF SPACE TRAVEL

- Aliens
- Alien overlords
- Spaceships
- Exceptionally shiny onesies
- Robots and humans working together to fight aliens
- Robots, humans and aliens working together to fight space monsters
- Socialist space-exploration utopia
- Interstellar politics—more boring than regular politics
- Hot alien babes
- Horrific genital injuries incurred from incompatible hot alien babes

- This guide
- Laser Boomstick
- Pressure suit
- Towel
- Spaceship
- Space-worthy time machine
- Backup space-worthy time machine battery
- Backup space-worthy time machine

INTRODUCTION

Finally, intrepid time traveler: space. This is where all those cool things you've been reading about in science fiction novels all your life finally take place: the aliens invade, they subjugate (and occasionally snack on) humanity, humanity revolts, everyone becomes friends and they tool around The Universe together, occasionally partaking in hijinks.

Before that can happen, though, there is the inevitable war between Earth and Mars. Well, not really Mars—they're just some guys who parked on Mars temporarily. Still, it makes for a good headline: EARTH VS. MARS!

No sooner does humanity throw off the bonds of slavery (and perfect, utopian communism—turns out it totally works if it's run by robots) than do aliens arrive for First Contact with humans—and subsequently enslave humans and robots alike. It's a tough occupation, lasting late into the second half of the millennium. Eventually, the Alien Overlords are overthrown by another group, the Alien Liberators, who aid human resistance fighters and who never again have a drink alongside a human without nudging him or her and saying in Alien, "We sure saved your asses in Worlds War II."

Thus, for the worries of the time traveler, arriving anytime between roughly 2323 and 2334 potentially throws you into an alien-human superwar, which is both horrific and as awesome as many video games and movies of the twentieth and twenty-first century suggest. Still, there are a lot of limbs melted off with lasers, being forced into despicable, tentacle-related sexual activities, and being served along with your fellow man as light, chunky bisque. But after that, it's a whole lot of space-faring, adventure-having fun. Suit up!

ALIENS

Know Your Aliens

1. X-Filian
2. Tentacled
3. Doctor Who
4. David Bowie
5. Floating Brain Alien
6. Xenomorph
7. Sexy Blue-skinned Alien Babe
8. Treacherous Lumpy Humanoid
9. Reptilian
10. Insectiod

Yup, they exist. It finally happens. And no, they don't arrive on our strange planet with no way of filtering out dangerous bacteria. They're also not allergic to the water that composes 70 percent of the Earth. And they certainly aren't afraid of nuclear weapons,[342] a technology they figured out roughly 2 million years ago.

But not all aliens are our evil betentacled overlords. Some are our good betentacled trade partners or multi-ocular interstellar peacekeepers. It is extremely rude to just show up in a post-occupation time of unparalleled tolerance, understanding and peace and start disintegrating immigrants and businessliens willy-nilly. Know your aliens:

Grays (X-Filians)

DESCRIPTION: These are the aliens you're used to seeing in popular culture from about the middle of the twentieth century. Three feet tall. Huge heads. Giant eyes. Silky soft skin. Balls of effin' steel—total bastards with psychic powers.

DURING THE OCCUPATION: You have two options—either drop to your knees like a sniveling coward and avoid getting psychically violated (or having your head exploded, as they're pretty quick on the draw with that one), or shoot first and shoot often whenever you see one of these things. The whole docile thing is just a ruse: They're pretty evil.

AFTER OCCUPATION: Turns out, despite some seriously bad blood between humans and grays, these guys control like 85 percent of the interstellar commerce in our sector of the galaxy.[343] In fact, the whole occupation thing is some kind of trade blockade to try and convince a larger galactic republic to take action or risk civil war—who knows. It's not something anyone understands, especially not George Lucas. Anyway, after a mere 220 years of occupation, the grays are ousted by that same galactic

---

342 For the record, fighting an invasive species with a nuke is like trying to get rid of an infestation by burning your house down. Congratulations: The ants have won.

343 As it happens, grays are not "evil," just capitalists. Tomato, tomahto, as it were.

federation-republic thingy, and Earth and X-Filia become the two biggest providers of entertainment and snack foods in Milky Way Spiral Arm 7. So no, you can't still shoot them.

Xenomorphs

DESCRIPTION: Eight feet tall, black, spindly, spiky, and fond of laying eggs in the chests of astronauts. These unthinking antlike killing machines are the bane of human colonies and the rest of sentient space-faring life. They enjoy heat and eating brains. Afraid of fire (but only just barely).

DURING OCCUPATION: Holy crap, you saw a Xenomorph on Earth? Oh my god oh my god oh my god. That's it, man—game over, man, game over!

AFTER OCCUPATION: AAAAAAAAAAAGHHHHH shoot it! Nuke the site from orbit—it's the only way to be sure without actually being sure. We don't care if the president and your grandma live there, that colony/city/home planet is toast.

Bowie

DESCRIPTION: There is no describing the Bowie. He simply is.

BEFORE OCCUPATION: Bow down and worship him as the space-man-god he is.

AFTER OCCUPATION: Bow down and worship him as the space-man-god he is, and thank him for being a hero of the human resistance against the X-Filian occupation.

Humanoid

DESCRIPTION: "Humanoid" is not technically a genus, but stems from the human tendency to think about everything from its own limited perspective. If an alien has two legs, can learn English, and could easily be confused with some guy in special effects makeup, it is lumped into the "Humanoid" category.[344]

DURING OCCUPATION: The hope was that by lumping all other upright-walking species together into a category named after a mutated version of us, humans would encourage some "humanity." Instead, humanity only offends the humanoids, and humanity is not afforded much humanity at all. Despite the rare instance of a successful courtship,[345] humans are generally looked down upon for their collective unwillingness to learn the difference between a Zertrand and a Klingon.[346]

AFTER OCCUPATION: As humans learn more about the vast array of alien species, and are humiliated by the desperate need for help from the Alien Liberators to save Earth from the Alien Occupation,[347] more care goes in to learning about the different alien cultures as well as recognizing the relative insignificance of mankind as a whole. There are some "good alien species" (like the sexy blue-skinned opera-singing Plavalagunas) and "bad alien species," but—as with humans—there are good and bad individuals in every species. Maybe the humanoids are just like humans after all. That would sure simplify things.

---

344 This makes intergalactic politics really hard, and intergalactic racism really easy.

345 See section "SURVIVAL GUIDE: SPACE TRAVEL: CAN I MATE WITH IT?"

346 Well, not a real Klingon. Those are made up. But these guys kinda look like Klingons and the human tongue can't really make the sounds necessary to say their actual species name, so we started calling them Klingons. This goes well until an ambassador brings a *Star Trek III* DVD to a meeting with the "Klingons" to give them a taste of human culture, and they murder him over the insult. So if you're going to say "Klingon," look over your shoulder first.

347 At which point WWII-related French jokes really no longer fly.

Phil Hornshaw & Nick Hurwitch

Can I Mate With It?

With an infinite number of new star systems and planets to explore come a nearly infinite number of new orifices and biological yearnings. That's right: Aliens have sex, too, and some of it quite casually. However, despite the innate sex appeal of that blue-skinned, triple-boobed space goddess, there are some serious questions you should be asking yourself before you just go around sticking/accepting tentacles into places they don't belong:

1. Is it considered bestiality?
2. Will I die or get the intergalactic VD?[348]
3. Is it, ya know . . . possible?

The answers to which are, in order:

1. This often depends on what languages your alien lover is capable of speaking. Low gutturals and sonic-boom voice boxes are typically bad signs. Fur and various appendages, however, don't necessarily make you a sexual deviant: After all, that's part of the fun of it, which is something we've only heard and don't actually know firsthand.
2. Most probably, yes. A good rule of thumb is: The more humanlike the better. And yet this same rule of thumb is a gaping, tooth-lined pit into which many a space-farer has unwittingly fallen. That's not a metaphor: Sometimes the most appealing mating partners based on external appearances are those most likely to contain rows of razor-sharp teeth, plant eggs inside you, or bestow unto you some horrific, drippy moon disease. Conversely, some non-humanoid, non-sexy aliens (like the FBAs) are the least sexually violent and also possess the least volatile insides.

---

348 There are some pretty horrific sexually transmitted diseases out there. Also, among some species, sex includes eye contact. Goggles are your friend.

3. As usual, that depends on your definition of "possible." Presuming you've "looked under the skirt" (to identify presence of razor teeth/sex tentacles) and are wearing adequate protection (to prevent becoming a living incubator for some demon spawn alien baby[349]), you can have sex with just about anything assuming the parties are willing, sober, and of age based on intergalactic legal standards and cultural norms.

However, if by "possible" you mean "baby makin'":

Successful crossbreeding with another species on Earth is difficult enough; nevermind doing so with a life form that possesses a completely disparate genetic evolution.[350] It's like trying to stick a round peg in a square hole. Over and over. And over.

Successful crossbreeding with an alien species usually doesn't work out as simply as "Aw, he has my eyes, and your ability to lift spacerocks with your mind." The results can sometimes be horrific, if not Space Circus Sideshow freaky. We're talking an arm here, a spikey pincer claw there, and extra brains where the toes should go.[351]

So remember: No matter how few members of your preferred sexual gravitation you've seen in the last several light-years of endless, black space, always take the time to think it through and consult your physician before attempting to mate with an alien.

---

349 Other horrific things that can happen to you: important parts can rot off; you can find your insides liquefied and slurped out of you; your interstellar communicator could start ringing at all hours of the night, leading to long discussions about "commitment" and "all those other aliens out there on the galaxy."

350 Helium-based lubricants burn a great deal.

351 And you thought you had it bad growing up with a gap in your teeth.

COLONIZING

The colonial spirit! Despite our lack of acid blood and ability to see more than three dimensions at once, the human race remains a major player in the space-faring portion of history thanks to that squishy, slightly yellowed portion of the soul for which we are well known.

We make great leaders, better lackeys, and we are willing to live just about anywhere so long as the tap water runs a shade lighter than the color of our excrement.

This is important for you to know because, even if you don't plan on settling down on some lava moon in the Kprulu Sector, a vast majority of "space missions" are a direct result of colonies. Check it out:

- Supply drops
- New colonizer drops
- Resource pickups
- Supply drop heists
- Colonizer heists
- Resource steals

Being familiar with the colonies will at the very least make the blinding complexity of starmaps somewhat less foreboding.

Using Trade Routes to Your Advantage

Some helpful tips:
1. Trade routes are typically named after the colonies at the beginning and the end of the route. The Earth-[Star Cluster 32] Route. The Tatooweenie-Endowindow Route. The Titan-Anuus[352] Route. If

---

352 Pronounced "Ann-*you*-us."

you know the name of the route, you know where you're (eventually) going.

2. For some reason, most planets are only well known for one or two resources and a single climate. If there's desert, it's all desert. If it's a forest planet, it's covered in forest. Water planets—you get the idea. This makes figuring out where you want to go as easy as picking out your favorite flavor of Skittle.[353]

3. You can reverse-engineer the ubiquitous nature of all planets into helping you gather resources. For example, if you are in desperate need of an umbrella, you need only to locate the trade route that ends at Soggankles, the Rains All the Goddamn Time planet. Any ship bringing supplies or new colonizers to Soggankles is sure to have an umbrella on board.

Living on a Colony

Unless you've committed some sort of heinous crime and need to lay low for a while,[354] you probably don't want to live on a colony. It's the small-town, single-restaurant, everyone-knows-your-every-foible kind of lifestyle, without the promise of escape to a larger town where your dreams may one day be fulfilled/crushed.

The governments and corporations in charge of establishing colonies are notorious for forgetting them. Look, it's about cost-benefit analysis. The cost is some civilian transports, the occasional terra-forming or ozone enhancement, and a bunch of castoffs without strong ties to family. The benefit is that only one of every dozen or so colonies needs to be profitable for them to make a killing.

---

353 Turns out Earth is unique after all.

354 In which case, you should just time travel away from that time.

Those are your odds of seeing routine supply drops: one in twelve. You have better odds of surviving black hole time travel.[355]

And even if you are that lucky one in twelve, we suggest you get real familiar with mining. Because that is all there is to do in your new space colony life: mine. Mine and gin rummy. Mine, gin rummy, and record a video log for the rescuers to find after they receive your distress beacon. But that's it.

Compounding your scant odds of survival are the detailed investigations and tests your overseeing government/corporation performs before sending in innocent volunteers. That is to say: They don't do any. In the race to grab up all of the valuable resources in the galaxy before competitor X, playing the numbers game is far more efficient than paying for years of "research" by "certified scientists" who would then probably make a "recommendation" to "not colonize the planet" due to the "inordinate amount of Xenomorphs" that keep "planting eggs in the research interns' chests."

If you do wind up on a colony, take the first tin can off that rock and don't think you can afford to be above certain types of "favors" to do so. This is your life we're talking about, and there's nothing dignified about dying of sodium-sulphide asphyxiation on the thirteenth moon of Anuus.

SPACE MARINES

The worst part about space colonies and their various trade routes is—what else—their fascist governments. You would think that a jurisdiction roughly $10^{48999534}$ the size of Los Angeles County would free up the clamps a bit, but they still manage to keep their proverbial space johnsons in everyone's business.

---

355 Hyperbole. Your odds of surviving black hole time travel are still considerably lower.

WHAT YOU NEED TO KNOW ABOUT SPACE MARINES:

- These guys are so badass, they smoke inside their spacesuits.
- Their spacesuits are not only good for breathing in space, but also for kicking ass in space. Made of a lightweight metal alloy, the suits double as body armor. Peeing inside of them is actually encouraged.
- They don't give a damn about how badly you need that oxygen to survive or your pansy-ass time travel woe-is-me bullshit. Don't cry to a Space Marine: he will as soon flush you out the air lock as ask you how tall you are and make off-the-cuff remarks about being unaware shit was stacked as high as your answer.
- They never take off their space suits. This makes the jokes easier (e.g., "What happened to the Space Marine who farted in his spacesuit? Answer: He breathed his own fart.") and getting your ass beat much easier if you tell any of those jokes within earshot.
- Space Marines are really just depressed. You'll notice they're never happy with their present situation: "I gotta get off this rock," "I gotta get off this tin can," "I gotta get off this standard issue marine gruel," "I gotta get out of this chickenshit outfit," etc.
- All Space Marines have a sordid past. Their planet/colony/battalion was likely destroyed by alien scum, probably due to the clerical error of a pencil-dick like you.

WHEN FIGHTING SPACE MARINES:

- Due to the bulky nature of their space helmets, the peripheral vision of Space Marines is extremely limited. Use the element of surprise or the element of slipping out the back escape pod.

Phil Hornshaw & Nick Hurwitch

Suggested foreign bodies and potential points of entry

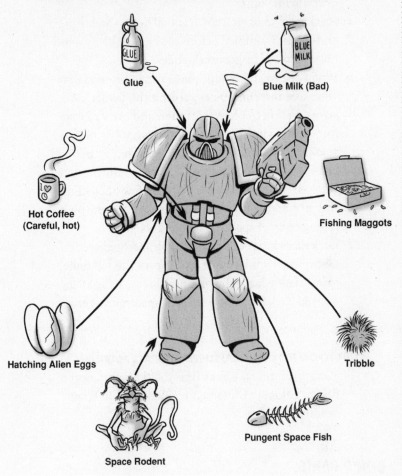

Glue

Blue Milk (Bad)

Hot Coffee
(Careful, hot)

Fishing Maggots

Hatching Alien Eggs

Tribble

Space Rodent

Pungent Space Fish

- Put something in their spacesuits. Glue and hot coffee are particularly effective, as are live space rodents. This will require you to wait until they lift their spacevisors.
- Power in numbers: Space Marines never travel alone and are always armed. If there are more of you with more or bigger guns, you'll at least delay the massacre

until their superior training and battle senses overwhelm you.

- Steal their smokes. They'll get all heavy-breathing and clammy-skinned before you can say, "Hey, asshole: I stole your gorram smokes."
- When dealing with multiple Marines, pretend that you have information regarding the leader's destroyed planet/colony/battalion and offer to take him there. Once he's separated from his chain-smoking, roughneck battalion, you'll have a better chance of befriending him and more time to get yourself out of this mess. Throw a few aliens into the mix and you might just have your very own ragtag crew of space-faring do-gooders.
- Lock them in a room with a couple facehugger Xenomorphs. It's inhumane, but it gets the job done. Immediately leave the premises before baby aliens pay you back for being a soulless spacemurderer.

WHAT TO DO IF YOU'RE CAPTURED BY SPACE MARINES:
- Congratulations! Looks like you'll be among the first to colonize Unnamed Ice Planet 87-B. Happy mining.

DON'T PANIC

Locate your towel.[356] Have it with you. Assess the situation and think clearly.

---

356 See also "SURVIVAL GUIDE: SPACE TRAVEL: TOWEL."

EARTH

Earth has seen better days. Namely, during Utopia. Ever since mankind used up most of its natural resources (which hurt a bit), the Robot Uprising (which was kind of a wash), centuries of robot-controlled soil (which actually helped a great deal), and the subsequent Alien Invasion (which hurt the most, but only because we nuked our own planet), the Earth's only major exports are entertainment, snack foods, saltwater and species lacking telekinetic powers.

Luckily, though, Earth's biggest problem—overpopulation—is kept in check by Space: the Second to Last Frontier.[357] Our pioneering spirit kicks in[358] and once again we get to work sending our least desirable residents to start colonies in unimaginable living conditions and spreading disease the universe over.[359]

Some important Earth-related things of note:

APE PEOPLE

If you travel to Earth any time after 4783, avoid The Island Formerly Known As Manhattan. The animal rights movement gets a little out of hand around the turn of the thirtieth century and, well, some humans successfully breed with apes. Luckily, the super-strong, incurably violent apes are also horrible swimmers

---

357 The Final Frontier being organic gardening. No! It's time travel! God, the book is almost over and you're still not getting it? Start over. Page 1. Go on, we're not kidding.

358 See also "SURVIVAL GUIDE: SPACE TRAVEL: COLONIZING."

359 Humanity also finds an incredible niche market as explorers, colonists and enforcers. The people of Earth are hired by most other alien species to do the kind of space-faring crap we've always dreamed about but no thinking, self-respecting sentient life-form would get excited about. We fling ourselves into black holes, land on planets to see if there's life there that will kill us, and scoop up whole species of cute, simple life-forms to feed other members of our Galactic Federation. It's at once a sweet gig and a sad, sad existence.

and struggle to operate most hovercrafts, so quarantine only costs the Unified Earth Government Council one of the five burroughs and several airdrops of kittens.

MORLOCKS

Once the Damn Dirty Apes do learn to swim, however, they accidentally go the wrong way and eventually end up in London, smelling about as awful as you can imagine a pack of soaking wet monkey-people would. The Brits attempt to reason with the apes, but their thick accents are simply too difficult for the apes to parse. The Apes respond with violence and the cost to the Unified Earth Government Council this time is the island nation of England.

Something about the drinking water or the way all the cars face the wrong way causes some peculiar evolutions in the DDAs, including blue skin and white hair. Their violent tendencies, however, remain intact.

Other than those two ape-related mishaps, though, Earth really starts to round third. Humanity remains active in politics and extracurricular activities, and some of us are even considered adequate space-faring companions by our humanoid cousins.

SPACE

The arrival of aliens following the Robot Wars signaled the true beginning of humanity's time as a space-faring race. The occupation and subsequent liberation brought with it humanity's induction into a galactic civilization, filled with incredible sights, new chances for exploration and trade, and some crazy-ass porn. Like seriously messed up.

Now that you're a space adventurer, there are a few new items to consider for general safety that haven't come up in past eras.

Phil Hornshaw & Nick Hurwitch

We've detailed a few things you should keep in mind about your new environment:

PRESSURE SUITS

Important for comfortably dealing with existence in the vacuum of space, as well as the cold-as-balls-ness of space and the airless suckfest of space. Keeping this suit sealed and supplied with heat and oxygen is key to your survival during extra-vehicular activity, during explosive decompression events (see below), and when fighting in some zero-gee circumstances. Bring a pressure suit, and bring a backup.

(Explosive) Decompression

"Decompression" refers to the escape of atmosphere from your spacecraft into space, resulting in a sudden shift in pressure. "Explosive decompression" refers to decompression that is so violent and fast, it creates an explosion in the hull of your ship.

- **IT'S NOT ABOUT YOU EXPLODING WITHOUT A PRESSURE SUIT WHEN SUCKED OUT INTO SPACE:** That doesn't happen. If you're exposed to a vacuum, you will not explode from the change in pressure within and without your body; nor will your eyes bug out of your head.
- **YOUR BLOOD WILL BOIL, THOUGH:** Near the surface of your skin. Other parts of you will bloat painfully.
- **YOU'LL ACTUALLY SUFFOCATE:** But, luckily, not before you pass out. You can survive in a vacuum for roughly fifteen seconds to a minute, but you'll likely pass out within the first twenty seconds or so. Quick action when exposed to a vacuum is key to non-death.
- **EXHALE OUT OF YOUR LUNGS AND DO NOT HOLD YOUR BREATH:** If you're attached to your lungs at all, you'll want to empty them of air when you enter a vacuum, or the pressure difference could make them pop.

- **YOUR EYES WILL PROBABLY BLEED AND GET NASTY, TOO:** Time to harvest that clone of you.[360]
- **GET INSIDE AND SEEK MEDICAL ATTENTION:** And you might not be permanently brain damaged, injured, paralyzed, or stuck wearing an iron lung attached to your chest.

ZERO GRAVITY

It can be a bitch. While zero-gee seems like fun at first, it has its own issues. For one, peeing is really tough when the pee just shoots out wherever instead of falling into the space-bowl. It turns into scary globule bombs that fly through the air, threatening to find their way into your mouth.

There's also the issue of atrophy—that is, your muscles will become weak and get gross and spongy if you don't use them, and the lack of gravity forcing you to propel yourself against its tyrannical grasp will cause you to turn into a scrawny, squishy shell of a man or woman. Upon returning to a place with real gravity, you could collapse into a humanlike jelly. Just, you know, FYI.

You're going to be spending a lot of time in zero-gee environments, so you had better get used to it.

Fighting in Zero-Gee

Note that without gravity to give you momentum or plant you on a surface, you can quickly find yourself free-floating into oblivion, totally defenseless.

MAGNETIC BOOTS: They're a good idea. Standard equipment on most extra-vehicular activity pressure suits. You can always just ~~staple~~[361] glue a big magnet to your pressure suit shoes, which

---

360 See section "SURVIVAL GUIDE: ROBOTS: CLONING."

361 Don't staple *anything* to your pressure suit. You won't be happy about it later.

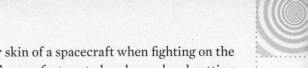

helps to grip the outer skin of a spacecraft when fighting on the hull. Drawbacks: With your feet rooted and your head getting whaled on, magnetic boots can turn you into one of those inflatable punching clowns, with considerably less merriment.

MAKING YOURSELF A MISSILE: You can always just kick off of something in the direction of your enemy, but the problem with this is, while you become a deadly man-missile, you're only able to go shooting off in one direction: go off half-cocked and you might miss your target entirely and wind up shooting your astronaut-missile self right into a sticky situation. Time and aim your explosive attack carefully.

FLAILING BALL OF FURY: Sometimes, you may just find yourself floating free near your target, momentum carrying you. This is one time that flailing like a seizure-stricken sissy is to your advantage: You become a furious flying ball of pain, increasing your chances of hitting your opponent without needing to see him or her. At the very least you'll bounce off something and change direction.

Bad Sex

Low gravity seems like it'd be useful for copulation—after all, without having to do things like support one's own body weight, one needn't worry that one has eaten one too many chicken wings and has begun to crush one's partner. Unfortunately, while it's true that fornication in space requires less manual labor in many ways, it's also as awkward as it is dangerous. Without the benefit of gravity, there is a great deal of . . . slipperiness . . . as well as strange applications of force and momentum. One minute you think you're giving your partner a little warp 9, the next moment you're bouncing off an electrical panel with the skin singed off your ass.

If you're going to insist on trying to do it in space, try an enclosed bed and "safety" straps.

Big Ships

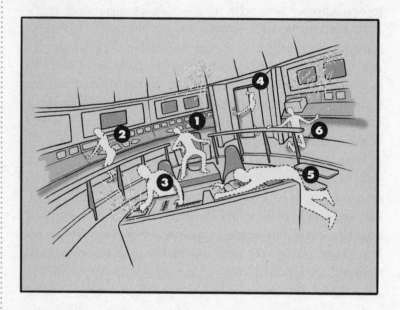

1. YOU—that's right. You get the comfy chair. Someone else might decide they want the comfy chair, too, so watch your ass.

2. XENOLINGUIST: Someone to help you communicate with any alien. Helps if he or she is sexy.

3. YOUR FIRST MATE: Someone you can trust, but will also challenge you on tough decisions. Helps if he or she is from a region of space where you have less political clout.

4. MECHANIC: You're on a light-speed-capable tin can. Things are going to break. Bonus points for mechanics with extra heads/arms.

5. PILOT: Someone's gotta fly this boat. Hope they know where they're going. If you're having a good year, recruit a navigator too.

6. DOCTOR: It won't be long before you come down with a rare space disease, more than likely due to your own idiocy. A resident doctor can help delay the inevitable. But don't expect much more from them, dammit, they're a doctor.

Big ships are controlled from the bridge. Despite the number of people and aliens this seems to require, driving one is shockingly easy. Space is primarily filled with empty nothingness, and the odds of you actually hitting something are damn near nonexistent.[362] The biggest objects, like stars and planets, will, at worst, pull you toward orbit, and any ships or space colonies with a dock shouldn't come anywhere near your hull: Your ship has transport pods for that, and even Jed down in janitorial can pilot one of those. Avoid comets, asteroid storms and warping into the center of a star and you'll be peachy.

Fighter Ships

Piloting a space fighter is not unlike piloting a fighter jet, with the exception of hypersensitive controls (due to a lack of friction) and, once again, the vast emptiness of space. This means that you'll have to manually stop maneuvers like the "barrel roll" or the "Flying V" or your ship will just continue the action until you're out of communication range, vomit out your insides, or shot.

However, the odds of you being finished off by an enemy starfighter (or vice versa) are considerably low for a number of reasons:

LACK OF DEPTH PERCEPTION IN SPACE—Without things like the ground and the sky to keep your surroundings in perspective, aiming becomes a bit of a crapshoot. Your best bet is to just hold down the trigger.

SIZE OF SPACE—"Battlefields" can now be as vast as entire star systems. At best, you're fighting in the shadow of a moon or moonlike space base. You could fly for hours—in the midst of battle—without having anything to shoot at. And with the lack of friction, everyone's going so fast that they might be gone be-

---

362 If you do hit something, fire your navigator and shoot your pilot: It's about time someone took control of this damn ship.

fore you've identified them as the enemy or shot down your commanding officer in a panic.

THERE'S NOWHERE TO CRASH IN SPACE—Let's get this out of your head: Ships don't just explode when shot. Who would ride around in something like that? Certainly not us. Most jet pilots only die once they crash into the ground (unless they really take a missile straight up the tailpipe). In space, you just kind of spin around until you're saved, forgotten, or become excruciatingly bored.

Your best bet in a starfight is the element of surprise. Because in space, no one can hear you gun down his or her wingman. Really: Other than your ship's communicator, which goes down every time it's slapped around by invisible space radiation, there is no sound. At all.[363]

TIME MACHINE REPAIR

So you're sitting in the Era of Space Travel, with aliens all about you and possibly a ragtag team of robots, humans and aliens headed off on Final Frontier, or at least Second-to-Last Frontier, adventures, and you're saying to yourself, "Gee, I have no idea how to fix my time machine."

Is that really what's happening right now?

BAD NEWS: If you're worried about how to repair the time machine and you're in the era of space travel, chances are pretty good you're in space. And if your spacecraft/time machine is broken, you're probably already dead.[364]

RETROFIT YOUR SPACESHIP INTO A TIME MACHINE: It's really not very difficult. You already have the spaceship. And you al-

---

363 Which puts a real damper on "Pull my finger" and other manners of flatulence humor.

364 See section "SURVIVALGUIDE: SPACE TRAVEL: PRESSURE SUITS, (Explosive) Decompression." That's probably what happened to you.

ready know how to make a time machine—how else would you get on a spaceship in the future? It's common knowledge at this point. Ask an eighth grader.

IF ALL ELSE FAILS: Use faster-than-light travel and relativity to escape into the FUTURE![365]

TOWEL

Bring one of these highly useful objects, and always know where it is.[366]

---

365 See section in Chapter 3 "Time Machines: Building Them and Inevitably Destroying Them for the Good of Humanity," "Relativity."

366 See also "SURVIVAL GUIDE: SPACE TRAVEL: DON'T PANIC."

SURVIVING IN TIME: THE END—OF TIME

13501 CE–?????

(Destruction of Earth to the Restaurant at the End of the Universe)

HOW TO KNOW IF YOU'RE AT THE END OF TIME
- Nothingness
- Blackness
- Old man
- Single street lamp
- Hallucinogenic experiences

WHAT YOU SHOULD BRING
- This guide
- It probably doesn't matter what else you bring

THE END OF THE ROAD

As with backward travel, during which getting overzealous with the chronometer can land you in an era predating oxygen, atmosphere, or the Earth itself, you should be cautious about how far into the future you're willing to tread.

We don't want to spoil anything, but Earth won't be around forever. And Earth especially won't be around after that pact with the Cthuliens goes south. As for space, well—space gets weird. Solar systems drift, stars die, black holes form: Keeping it all straight without some sort of Home Base to return to is nearly impossible.

The adjustment to life post-Earth and post–Homo sapiens is not one we have ever heard of any individual successfully making. Besides: Do you really want to know what waits beyond the far reaches of spacetime? Science indicates that it might simply be Nothingness, the Bigger Bang, or like that part in *2001* where Dave turns into a wrinkly old man and then a fetus really, really quickly. And what's with that room he's in? Is that the end of spacetime?

Perhaps the room is enlightenment: a meeting with God him- or her- or itself. Or perhaps, as here at QUAN+UM we are inclined to believe, it's just another really, really interesting way to get yourself killed.

SOME FUN FACTS ABOUT QUAN+UM

- We do NOT [officially] accept government funding.
- We are NOT [technically] a for-profit organization.
- We do NOT [strictly speaking] trust the banking system.
- We PREFER the colors purple and orange [but never together].
- We can NOT [legally] accept credit for forming the most awesome team of time-fighting travelers ever assembled, which comprises such time travel greats as Alexander Hamilton, Delbridge Langdon III, and Winona Ryder.[367]

---

367 But that was totally us.

INTERNMENTSHIP OPPORTUNITIES

CONGRATULATIONS! YOU'VE FINISHED reading the guide front to back and are well on your way to time travel expertise, or otherwise were one of those kids who dumped out all the cereal in the box just to get to the prize faster. Either way, we admire your wanton enthusiasm.

Now you're probably wondering to yourself, or aloud to this non-sentient book, "When the hell am I going to get to time travel already?" Well, literate compatriot, we're glad that you asked.

Here at QUAN+UM we operate using a rigid, top-down, high-work, low-reward corporate ladder structure. This means that should you decide to pursue temporal dislocation with us, you get to begin your would-be saga of rich adventures the same way that each of us did: in QUAN+UM WEDGIE (Wormhole Educational Development & Guided Internship Experience), time travel's premier internship program.

Some of the exciting opportunities that await you as a QUAN+UM intern:

HANDS-ON LEARNING—Obtain an advanced degree in floor mopping and timestream cleansing.

CULINARY EXPERTISE—Learn the food-preparation secrets hidden in human history. Go on food runs to the far reaches of Earth's many epochs.

LEARN TO USE A FIREARM—The best teacher is practical experience. It could save your life!

BE A PART OF HISTORY—But hopefully not in the past. That sort of thing is forbidden.

BECOME A TRUE TIME TRAVEL PIONEER—Participate in experimentation and reconnaissance critics and observers have called "thrilling" and "untested."

FREE ROOM AND BOARD—That's right: QUAN+UM will provide you with your very own on-site high security cell and as many of our leftovers as you can eat.

"WELL THAT ALL SOUNDS SWELL. HOW DO I SIGN UP?"

Believe it or not, the authors of this book and our affiliates do have office space in a real, physical location.[368]

Unfortunately, we are not able to disclose the coordinates of said physical location, as even in an era when time travel is known to be possible, those who practice it are looked upon with an air of "You're going to get us all fucking killed, guys." But if you do manage to figure out which building is ours, feel free to swing by; that's at least worth a proverbial time traveler fist bump or comparable barbarian-influenced physical greeting.

---

368 Though to be fair, we rent.

Plus, we'll get you started on the internship ~~liability waivers~~ paperwork right away.

HQ is also a pretty good place to duck your head once the robots start attacking because, well, we knew about it.[369] Though our eighteen stories and several thousand square feet of lecture spaces, laboratories and handicap-accessible bathrooms crumble under the might of The Jawbreaker[370] like so many other structures in **[redacted]**, our underground facility is heavily stocked with the provisions necessary to continue our research, aid in the occasional one-way portal launch of a reprogrammed enemy robot, and keep paradox-waiting-to-happen amateurs like yourself out of trouble.

If you get cold feet before signing ~~your life away~~ up, be sure to check out the gift shop and poke around the restricted areas to say hello to a certain lab coat–wearing duo of pure, unadulterated Science.[371]

"BUT I'M UNDERAGE, UNEMPLOYED, OR DON'T HAVE MY OWN TRANSPORTATION."

If signing up in person isn't an option for you, we retroactively began taking applications online in 2012 at www.thetimetravelguide .com. Head there now and start your time travel dreams today.

Just bear in mind that, as usual, you should keep your expectations reasonably low. Time travel is a burgeoning field and internship cells are limited. If you are not one of the lucky few accepted into QUAN+UM WEDGIE, fear not. You still hold the power of time travel in your hand. No, not that hand, the one

---

369 A short-lived "I told ya so" to all the haters. Short-lived due to all of those haters dying in the Robotacalypse.

370 A robot originally designed to dig through the center of the Earth for the first Manhattan-to-Beijing subway line, later re-purposed to dig through mounds of sniveling humans. See section "SURVIVAL GUIDE: ROBOTS: ROBOT MODELS (COMMON), Tunnel-Digging."

371 No free autographs.

with the book in it. So long as you follow *The Time Traveler's Guide to Time Travel*'s suggestions and loose guidelines, you'll probably live to tell about it. Not to mention, we'll be less inclined to hunt you down and erase you from the timeline for the good of humanity. Which is something we do. Often.

Happy travels.

ABOUT THE AUTHORS

CHRONO-DISPLACEMENT MOLECULAR Reassembly Psychosis Specialist and Test Subject Hostage Negotiator **PHIL HORNSHAW** moonlights as a technology blogger when he hangs up his lab coat. A sometimes fanatical consumer of television and film as well as part-time payer of attention to Science, he has known a fascination with time travel and its horrific effects on the human mind and body all his life. He currently lives in Los Angeles, alternating between conducting chrono-displacement experiments on Nick Hurwitch and writing about playing video games on his cellular telephone.

NICK HURWITCH read his first book on theoretical physics at age twelve and ever since, has been both fascinated by the Universe's mysteries and content to let someone else figure that crap out. He is now a writer and filmmaker living in early-twenty-first-century Los Angeles. The imagination is his metaphysical realm of choice and although he technically lacks the credentials of a time scientist, he is happy to lend said imagination to the world of quantum physics. He hopes that publishing this book will mean no future time travelers will have to relive his awful experience in Budapest, 1903.